Science Left Behind

SCIENCE LEFT BEHIND

Feel-Good Fallacies
and the Rise of the Anti-Scientific Left

ALEX B. BEREZOW · HANK CAMPBELL

PUBLICAFFAIRS
New York

Published in the United States by PublicAffairs™, a Member of the Perseus Books Group

All rights reserved.
Printed in the United States of America.

PublicAffairs books are available at special discounts for bulk purchases in the U.S. by corporations, institutions, and other organizations. For more information, please contact the Special Markets Department at the Perseus Books Group, 2300 Chestnut Street, Suite 200, Philadelphia, PA 19103, call (800) 810-4145, ext. 5000, or e-mail special.markets@perseusbooks.com.

Library of Congress Cataloging-in-Publication Data

Berezow, Alex B.
   Science left behind : feel-good fallacies and the rise of the anti-scientific left / Alex Berezow, Hank Campbell.—1st ed.
      p. cm.
   Includes bibliographical references and index.
   ISBN 978-1-61039-164-1 (hardback)—ISBN 978-1-61039-165-8 (electronic) 1. Science—Political aspects—United States. 2. Science and state—United States. I. Campbell, Hank. II. Title.

Q175.52.U5B47 2012

303.48'30973—dc23

2012012544

First Edition

10 9 8 7 6 5 4 3 2 1

*For my wife, Ania, without whom all my
endeavors are without meaning*

    —ABB

*To Kim, who has always been supportive, even six
years ago when I decided I wanted to write science
on the Internet. Joe, Courtney, Paige, Colin, Cormac—
you got your names in a book*

    —HJC

CONTENTS

INTRODUCTION

The Progressive War on Spoons

IN 2007, FRESH OFF A 2006 VICTORY in both chambers of Congress, Democrats set out to boldly fulfill their campaign promises to make America more sustainable—by replacing utensils in the U.S. Capitol building. They called it their "Green the Capitol" Initiative. With this initiative, the Democrats planned to make the Capitol building itself a model of sustainability and set an example for us all. They wanted to do just what they thought America needed; namely, update some lightbulbs and replace bathroom fixtures—basically a progressive "honey do" list—but perhaps most importantly for their environmentalist constituencies, they took the giant step of "greening" the congressional cafeteria.

Cost was no object. Good thing, too.

Contrary to what the public often thinks, politicians and their staff members work hard. (This is not to say that they pass much meaningful legislation—besides drafting bills no one can actually read, most of their hard work boils down to finding new ways to take down the opposition.) Staffers put in long hours, and meal breaks are frequently to go. In 2006, that meant a heavy reliance on Styrofoam and plastic utensils. To the new Democratic majority of 2007, this meant shocking environmental waste.

Their solution: new, greener forks and spoons that were better than plastic because they could be composted.

They claimed science was on their side. Composting advocates have insisted that composting on a large scale would save money and be better for the environment. Never mind that the Capitol grounds do not really have room for piles of fermenting garbage; evidence is often first to fall in the culture war on science. Armed with a similar majority in the Senate (and an armful of new utensils), the Democrats issued a new decree: from now on, the cafeteria in the Capitol would offer takeout utensils made from corn and containers made from sugarcane.

The result was a miracle of sustainability, at least according to internal reports by the Democrat-controlled House. The program claimed to have kept 650 tons of waste out of landfills between 2007 and 2010.[1]

Greening the Capitol looked terrific on paper. The only problem: the new green replacements were actually worse for the environment and costlier as well.[2] The spoons melted in soup, so people had to use more than one just to get through lunch. The knives could barely cut butter without breaking. Hard as it was for the utensils to make it through a meal intact, it was even harder for them to actually biodegrade afterward. Instead of being easily composted, the utensils and containers had to be processed in a special pulper and then driven to Maryland in giant emission-belching trucks.

The House Inspector General ordered a new, independent analysis in 2010 and found that the actual emissions savings in return for all that money and inconvenience was equivalent to removing one automobile from use[3]—at a cost of $475,000 per year. Wary of disappointing their environmentally progressive constituents, Democrats waited out the clock until the Republicans regained control of the House in 2011, at which point they suggested that Republicans kill the program. Republican Representative Dan Lungren of California, the new chairman of the House Administration Committee, listened to the advice of his Democratic predecessor, Representative Robert Brady of Pennsylvania, and instructed the cafeteria to revert to using utensils and containers that actually worked.[4]

Who was blamed for being an enemy of the environment for that decision? Progressive Nancy Pelosi tweeted that Republicans were the problem: "GOP brings back Styrofoam & ends composting—House will send 535 more tons to landfills,"[5] she claimed. Did progressives question her math? No, but they should have. The initial program supposedly saved 650 tons of waste in three years, or about 217 tons per year. But according to Pelosi, without the program, the House was going to generate 535 additional tons of waste per year. Her math simply didn't make sense.

This quirky anecdote is very illuminating: if an environmental story is being framed about people on the right, anything goes, but if the numbers don't add up for a progressive cause, the bad arithmetic is ignored. Representative Mike Honda (D-CA) even went a step beyond Pelosi, contending that an end to the composting program and a return to Styrofoam would *cause a spike in cancer*. "These cancer-causing cups leach toxic chemicals, which threaten our health, our reproductive systems, and our environment as the nation's 5th largest creator of hazardous waste. To claim these cups are part of a cost-cutting measure is completely disingenuous."[6]

The right is not more anti-science than the left; it just has terrible public relations. Progressives have mastered feel-good fallacies, and they've become so proficient at it that they are able to convince and sometimes bully the scientific community into playing along.

Feel-good fallacies designed just to win hearts (not minds) and the kooky left's hijacking of science and medicine with dangerous implications are what this book is all about. Just like the House Democrats' corn spoons and sugarcane cups, these policies often end up doing little and leaving behind a big mess. Worst of all, anyone who questions them is framed as anti-science.

If, while reading this introduction, you looked at your dishwasher and wondered why environmentally conscious progressives had not simply bought reusable utensils, the kind you use in your house every day, instead of wasting taxpayer money on a social experiment, then this book is for you. If you ever wondered why progressives have a strange fetish with alternative energy, then this book is for you. If you have ever wondered if organic food is nutritionally better as progressives claim, then this book is for you. If you ever wondered why progressives claim to support government policies "for the children" but then refuse to vaccinate them, then this book is for you. And that's just a small sampling of the many topics we cover. What we describe in these pages amounts to a progressive war on science and reason.

WE RECOGNIZE that even using the term "progressive" is going to alienate some of the public. Politics is divisive. Any book that deals with politics has an instant credibility problem: before anyone even opens it, they have a predetermined opinion about how they're going to react. To that end, we're going to lay our cards on the table so that you know exactly what you are getting and why this book needed to be written.

We love conservatives, we love liberals, and we love libertarians. Those groups are everywhere in American culture, and their diversity of thought is what fosters a healthy atmosphere that makes cutting-edge research possible.

We love conservatives because of their adherence to tradition and to the principles that have made the United States the most successful country on earth. We love liberals and libertarians because of their insistence on freedom. In particular, we favor liberalism in the classical sense, as philosopher-physician John Locke defined it—the pursuit of "liberty." We are radically liberal about scientific thought. Science should be far beyond the reach of agenda-driven politicians and activists. Science should be free to speak for itself, not be held hostage by partisan politics. And most importantly, science policy should be driven by data, not ideology. Therefore, in the classical sense, we are staunch science liberals.

Science rigorously tests hypotheses and theories using well-controlled experiments as a way to understand reality. Politics is often more about photographing your rival in a compromising situation and e-mailing it to CNN. (That's a lesson Representative Anthony Weiner learned the hard way.) Science lets the data speak for itself; politics is about spinning the data to score points against the other team. One pursues objective truth; the other, subjective truth. You can especially see this latter point in today's polarized political dialogue.

Yet in the midst of this subjective, polarized environment, science and politics have become increasingly intertwined. Today, the majority of basic research is through government grants.[7] Moreover, America leads the world in science. No one else comes close. Only the dangerous trend of politics infiltrating basic science puts this country's leadership in question. And of all of today's political philosophies, progressivism stands as the most pressing problem for science in our country.

Progressives, not liberals, are the ones you see in the headlines trying to replace scientific research with unscientific ideology. They are the ones banning sales of goldfish and creating no-win "games" like Gender Bias Bingo. A progressive atmosphere is simply incompatible with producing high-quality science. Though progressives try to claim common cause with liberals, today's progressive movement is more an umbrella of clustered, socially authoritarian interests. But science has little to do with progressives' pet beliefs and fad initiatives.

The purpose of this book is not to vindicate Republicans or Democrats or any political ideology. Okay, so we might poke fun at some people along the way. Mostly, they will be Democrats because very few progressives are Republican. (Honestly, how can you not make fun of a progressive professor who finds a way to justify having sex with animals? If you find that titillating, you'll be thrilled to read Chapter 8.) However, despite what some progressives will contend, the purpose of this book is not to demonize all progressives. We just want to demonize the loony ones. We recognize the tremendous value some progressive ideas have had in shaping American culture over the past century, and vilifying an entire philosophy based on the actions of radical ideologues would obviously be unfair, just as it is unfair when progressives do it to conservatives.

So we won't do that. The purpose of this book is instead to inform you about a disturbing trend among highly influential progressive activists who misinterpret, misrepresent, and abuse science to advance their ideological and political agendas. Though some progressives are pro-science, many within their ranks are not. They bogusly wave the banner of science while peddling pure mythology, and they are deaf and blind to all evidence to the contrary. It is our intention to call them out.

The conservative "sins" against science (e.g., ethical concerns about human embryonic stem cell research, skepticism about climate science, and fringe religious opposition to evolution) are widely reported and well known. Other books have already rehashed these themes, and during every election cycle the media aren't shy about reporting how scientifically ignorant some progressives in science claim conservatives are. We have nothing constructive to add to that part of the dialogue—except that the reverse phenomenon of progressive hostility toward science remains strangely underreported. And, in our opinion, this trend is just as, if not more, insidious and as relevant to the average person as the supposed Republican war on science.

Among many other examples, progressive activists such as Robert F. Kennedy Jr. have championed the unscientific anti-vaccine movement, confusing parents and causing a public health disaster. Meanwhile, even as animal research remains scientifically necessary, progressive groups advocate against it and, in some cases, have committed acts of violence against medical scientists. Instead of embracing technological progress, such as genetically modified crops, progressives have spread misinformation while peddling organic foods that are nothing more than unscientific scams. Progressives have waged war against academics who question their ideology, and they are opposed to sensible reforms in science education.

As we will show throughout the chapters that follow, techno-phobic tendencies, anti-corporate bias, and an obsession with relativism at the expense of empirical proof—all of which partially form the basis of the modern-day progressive movement—have produced a generation of progressives who reject scientific and technological advancement. Unreported by the media, these beliefs and their ramifications have remained largely unknown to the public. Until now.

Normally, conservatives who endorse policies that are scientifically unpopular are blasted by the media and scientific community. Yet progressives who do the same generally get a free pass. *It is the responsibility of the community of scientists, science journalists, and the American public to ensure that this free pass be revoked.*

And with so much already reported about the alleged "conservative war on science," it is time to present, as Paul Harvey famously said, the rest of the story.

Chapter One

WHAT'S A PROGRESSIVE?

They're Not Liberals, but They Think They Are

FEEL-GOOD FALLACIES about spoons made from corn are great for outrage radio, but compared to some other progressive myths, they are not really harmful. If only all progressive policies were so benign. It's when progressives venture into bizarre disinformation—just look at the parents who claim to care so much about their children they deny them vaccines for diseases that could cripple and kill them, or the food fanatics who think pasteurization is harmful but raw milk is safe because it is natural—that their kookiness becomes a lot less amusing.

So let's begin with the most basic question of all: Who are the people we're calling progressives? Generally, they're the kind of people who think that overpriced granola from Whole Foods is healthier and tastier. They're the people who buy "Terra Pass" bumper stickers to offset their cars' carbon emissions. And they're the sort of people whose beliefs allow them to feel morally superior to everybody else who disagrees—even if scientists are among those doing the disagreeing.

In general, progressives vote for Democrats, but not all Democrats are progressives, just as not all Republicans are conservatives. As you read this book, please keep in mind that a liberal

New York City unionized police officer has very little in common with a progressive San Francisco hippie who smokes organic weed and lives in a tree. Though they both tend to vote Democratic, they are not actually culturally or ideologically aligned.

Additionally, progressives have waged a largely successful smear campaign against conservatives, even going so far as to accuse the entire Republican Party of being anti-science. They spend a good portion of their time justifying many of their half-baked, pseudoscientific beliefs while simultaneously claiming that the Democratic Party is the true champion of science. Honestly, if they didn't do that, this book wouldn't be necessary. We don't have a dog in the never-ending political fight between Team Red and Team Blue. However, we are zealous foot soldiers for Team Science, and we will push back against unjustifiable attacks on other people, especially when those doing the attacking hold a plethora of anti-science beliefs themselves. We engage in this battle not on behalf of Republicans, but on behalf of good science.

We openly admit that some conservatives (and, by extension, some Republicans) have embraced anti-science positions, most famously on evolution and climate change. So have some progressives (and, by extension, some Democrats). This book isn't about Republicans and Democrats; it is about progressives. And we intend to demonstrate that if it is true that conservatives have declared a war on science, then progressives have declared Armageddon.

(Progressives vs. Liberals) vs. (Conservatives vs. Libertarians)

But, first, let's clear up some terminology. It would be too easy to say that the culprits are liberals. It would certainly sell more books. The word "liberal" has become a political buzzword, and it has been effectively used (usually by conservatives) to smear a lot of people on the other side. But that's precisely the problem: without further clarification, the word doesn't tell us much of anything these days. Instead, it means different things to different people, and some use that to their advantage. As a result, many people think that "progressive" and "liberal" are interchangeable labels, but they are not. Progressives like to think of themselves as liberals, but they really aren't. They're radically different.

A political party really is far less dictated by ideology than many voters like to think. Politics is messy and governed by expediency. Ideas and initiatives are just a means to a very simple end: getting elected. Though the matter rarely comes up except during election campaigns, Republicans often vote for Democratic initiatives and vice versa. Parties and their goals also change over time. Democrats weren't always liberals or the party of "equality," and Republicans weren't always conservative or obsessed with interventionist foreign policy. The old Southern Democrats of the late 1800s would be almost unrecognizable to the Bill Clintons, Jimmy Carters, or Barack Obamas of today. And Republican Richard Nixon had far more in common ideologically with Democrat John F. Kennedy than he had with fellow Republican Ronald Reagan. So to find the real source of these anti-scientific ideas, we must look beyond buzzwords and party names to the concepts at the heart of politics.

Sorting people's political ideas is a difficult task. But in 1971 American libertarian David Nolan attempted to redefine how we look at political identification by creating a quadrant approach— one that happens to be exceedingly helpful in finding the source of the ideas at the heart of today's anti-science advocates. Nolan's grid gets rid of the overly simplified left-right divide of Democrats versus Republicans by rating how economically and socially authoritarian each person is. His approach breaks people into four political camps: Libertarians, Liberals, Conservatives, and Progressives.

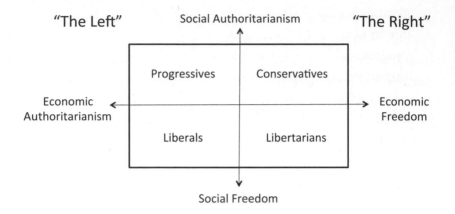

In the quadrant system figured here, conservatives and libertarians together constitute the "right" end of the American political spectrum. True libertarians are the people who seek freedom both economically and socially. They want government to be as irrelevant and laissez-faire as possible. Though they have made some concessions in modern times, most libertarians believe that government should be small, its primary purpose limited to defending our national security and fixing potholes in roads. They strongly believe in free markets, personal liberty, and individual responsibility.

Conservatives are similar to libertarians as far as economics is concerned, but they differ in their authoritarian approach to social issues. They seek to impose, and even legislate, moral standards on issues such as sexual conduct, drug usage, and abortion. For example, conservatives generally believe a person should have the right to purchase and smoke cigarettes, but not marijuana. In a two-party system like America, libertarians and conservatives primarily vote Republican.

On the left side are liberals and progressives. Like libertarians, liberals value social liberty, but they tend to favor greater government control of and intervention in the economy. In the abstract, they tend to favor higher taxes on the wealthy, more regulations on the marketplace, and social programs that redirect money to target social inequality.

Progressives likewise favor economic control but—and here is the important part—they are social authoritarians, much like conservatives. However, unlike conservatives, they are not concerned with banning "immoral" things like sex, drugs, and rock and roll. They instead seek dominion over issues such as the environment, food production, and education. They endorse bans on plastic grocery bags, McDonald's Happy Meal toys, and home schooling. They hold opinions that are not based on physical reality about how energy and development should work. And, most significantly, they claim that all of their beliefs are based on science— even when they aren't.

Thus, one of the primary goals in this book is to address the anti-scientific bias inherent in progressivism today. As we've said, when you read the word "progressive" in this book, you shouldn't substitute the word "liberal" or "Democrat"—they are not synonyms. Progressivism is a way of thinking that goes beyond party identification. Some Democrats are social conservatives. Some self-identified Republicans hold some progressive views. Progressivism as a political philosophy is more than a century old; in its earlier incarnation adherents of this philosophy included such figures as Teddy Roosevelt and Woodrow Wilson. Though progressivism has transformed greatly over the years, its contemporary expression follows in the footsteps of a long and storied historical tradition.

The Roots of the Progressive Movement

Before the turn of the twentieth century, big business ruled the American economy. We can complain about modern recessions now, but they are nothing compared to those that occurred in the nineteenth century, when there were depressions in 1807 and

1870 and panics in 1819, 1837, 1857, 1873, and 1893. Booms and busts destroyed lives and fortunes.[1] Wealth was consolidated in the hands of a few, but the concept of wealth had already begun to change with the spread of the Industrial Revolution. A nobility that had once valued clothing now saw a world where everyone could buy the same cotton only the wealthy once had. As wealth, and therefore education, increased, the rising middle class began to resent the power of the wealthy and the effects of a completely laissez-faire economy on poor people.

As a result, during the period between the beginning of the Gilded Age in 1877 and the 1920s, progressivism arose as a highly influential movement in American politics. Progressivism had two forces competing for dominance. One branch wanted responsible advancement into the future. Republicans like President Teddy Roosevelt and Representative John F. Lacey used government to create national parks and laid the foundation for modern conservation laws. The other branch served as a reaction to industry corruption and the perception of "impurity" in American society. That branch gave voice to the middle class's feeling that society needed to be reformed and its institutions regulated by government fiat. With this second branch arrived social authoritarianism—both the "good" kind, interested in initiatives like regulating impure food and preventing phony medicine from poisoning unwitting patients, and the "bad" kind, which advocated less defensible ideas like social Darwinism, eugenics, and the banning of alcohol.

For a time, progressivism made for good politics. Roosevelt was joined under the banner of "progressives" by Democrats including Woodrow Wilson and William Jennings Bryant. All of these men aimed to mobilize rationalism and science to promote "progress," just as their philosophy's name suggested.

Yet, just at the pinnacle of progressivism's rise, it became the victim of its own success. One of progressivism's signal achievements, Prohibition, focused on regulating personal conduct, like progressives do today, but they also instituted economic restrictions like the Smoot-Hawley Tariff Act, which raised tariffs for imports to historically high levels and was designed to promote purchase of domestic products. In effect, it actually made the Great Depression even worse.[2] Many progressives also believed in the promise of the racist pseudoscience called eugenics. The proponents of eugenics were a veritable Who's Who of the progressive movement. Names like Supreme Court justice Oliver Wendell Holmes, economist John Maynard Keynes, and Carl Brigham, inventor of the SAT. In *Mein Kampf*, Hitler praised the eugenics-based Immigration Restriction Act of 1924 because it not only banned entire ethnic groups; it banned people with hereditary illnesses too.[3] Many American progressives (but certainly not all) supported Hitler's eugenics policies in return, though claimed it was because they appreciated his science, not his autocratic style.[4] Regardless, the theoretical world gave way to the applied one, and Nazi Germany's policies sent the reputations of eugenics proponents into a tailspin and progressivism along with it. Progressivism in its original form effectively died.

It was not until the 1960s that the movement was reborn into something we would recognize today. Critics of academia had long been concerned about a gradual shift leftward, but their concerns about an increasing lack of political diversity mainly revolved around the humanities (as enunciated by William F. Buckley in his 1951 book *God and Man at Yale*). The publication of Rachel Carson's *Silent Spring* in 1962 was the real demarcation point when modern progressives left behind the scientific and über-rational legacy of the past. From that point on, they embraced radical environmentalism and other visions of a natural utopia.

Silent Spring used a lot of the same logic that anti-vaccine progressives use today: anecdotal evidence and dubious statistics coupled with generous doses of paranoia and very little science. Scientists such as Dr. Ira Baldwin, professor of agricultural bacteriology at the University of Wisconsin and later vice president emeritus and professor emeritus of the department, dismissed Carson's book in *Science* as a "prosecuting attorney's impassioned plea for action," not a book based on scientific data.[5]

Throughout the 1960s, society was in turmoil and incipient progressives found a new slate of causes to support. Ralph Nader's *Unsafe at Any Speed* documented carmakers' lack of interest in making their cars safe. The anti-nuclear movement also provided new targets for protest. "Population bomb" advocates like Paul Ehrlich lent a voice to progressive believers in neo-Malthusian ecology and drove a new interest in food and fertility. It's no surprise, given the tortured science legacy of progressives, that Dr. John Holdren, who now holds the most important science position in America (senior adviser to President Obama on science and technology issues, or "Science Czar"), wrote a book with the discredited Ehrlich in 1977 that discussed forced sterilization and mandatory abortion as methods of population control.[6]

By the modern era, super-rationalist progressives who once had held an almost religious belief in the power of science to create a utopian future had now largely left science behind. Lacking an emphasis on objective fact, and focused primarily on legislating ideology and fighting anything that disagreed with cherished ideas, progressives became as we know them today: unscientific, while claiming the mantle of modernity, denizens of a world where science is replaced by feel-good fallacies.

Progressive Mythology

This brings us to the anti-scientific school of progressive thought and activism that holds society in its grip today. Most current wrongheaded progressive ideas can be traced back to these four mythologies:

1. Everything natural is good.
2. Everything *un*natural is bad.
3. Unchecked science and progress will destroy us.
4. Science is only relative anyway.

Believers in the first progressive myth are everywhere. They shop at Whole Foods stores and believe that homeopathy and herbs are as good as actual medicine. Some even take jogs without shoes on because they think if ancient man did so, it must be correct—without recognizing that ancient man invented shoes for a reason.

If something comes from the earth, these people think, it *must* be healthy. Humans are products of nature, and therefore all we really need to be healthy can be found, unmodified, in nature. Of course, rattlesnakes come from nature, too, as does arsenic. And how about the entire process of agriculture? Plowing, sowing seeds, harvesting—none of that is, strictly speaking, *natural*. In order to cultivate food, farms destroy the existing environment. The truth is, the word "natural" has changed meanings—to progressives it now means anything that puts us in harmony with the earth. If you personally choose to believe in New Age philosophy and want to give holistic medicine a try, go ahead—but when the myth of the natural interferes with politics and science policy, we end up with dangerous outcomes.

The second myth follows easily from the first. If everything natural is good, then everything unnatural obviously must be bad. Believers of this myth contend that once humans interfere with something, it's no longer natural. Notice how humans, accepted by the first myth as natural, have suddenly lost all naturalness. Human intelligence and skill have likewise lost their wholesomeness. To progressives, the very idea of an unnatural substance, like a vaccine against a debilitating disease, becomes loathsome despite saving millions of lives. Natural gas becomes unnatural because progressives believe "fracking" is prohibitively dangerous and might even cause headaches[7] (though maybe acupuncture can cure those). Medicine, technology, and science itself increasingly become the enemy—a problematic opinion the consequences of which too many progressives take lightly.

The third myth is just the logical extension of the first two— science and technology are great, but only until they go too far. To a modern progressive, technology is inherently dangerous. That is why progressives oppose nuclear power and genetically modified food. Every other day, we hear how scary chemicals are causing cancer, even if little evidence exists to support the claim. Progressives demand that new technologies be proven safe—a scientifically impossible standard. Yet progressives consider themselves reasonable and educated people, part of a "reality-based community." In their minds, they stand in stark contrast to conservatives, who disbelieve climate change and evolution. But in reality, the criticisms they direct at conservatives also apply to them. While claiming to wholeheartedly trust in science and scientists, progressives do so only until scientific findings fail to uphold their cherished progressive values.

When that occurs, progressives accuse the scientists of transforming into shills for big industry or evil geniuses experimenting with world destruction. As soon as a cost-benefit analysis leads a corporation to do something that isn't green enough, businesses are accused of putting profits over people. As we explore in the book, this leads to policies like Europe's precautionary principle, putting the burden on corporations to prove that a food or chemical is 100 percent safe, as opposed to the status quo in America, which is that the burden is on the opposition to prove something unsafe. It's an innovation-killing policy. Additionally, the larger myth leads to important initiatives and worthy scientific endeavors being dogged by protests, often ending with them being shut down for reasons that violate true progress, economics, and even common sense.

What's behind this myth? Well, progressives truly believe that utopia is possible. Of course, it's their version of utopia, which stems from a sort of condescending, paternalistic conviction that they know what is good for the rest of us. Thus, any kind of rational cost-benefit analysis is anathema to them. Actual green energy is an unattainable goal for precisely this reason—as quickly as scientists come up with a new method for powering our society, progressives dismiss it as too unnatural and dangerous. Indeed, if progressives ran energy policy, most of us would sit in the dark—that is, unless we were filthy rich, like Al Gore, who can afford to plaster his house with inefficient solar panels.

The final myth is a crucial piece of the puzzle. Along with their faith in the natural, their horror at the *un*natural, and their suspicion of science and industry, today's progressives have one last core idea: everything is relative.

Postmodern relativism lingers in the background and binds together all of these myths. Progressives don't entirely buy that science has a unique claim on secular truth, like understanding the world according to natural laws, or that there are even any natural laws. For them, science is "just another opinion," an alternative worldview. A medical doctor's advice to "get a vaccine" holds no more weight than a local guru's instruction to meditate and drink herbal tea. In a different way, this same relativistic bent permits progressives to refuse to admit research that shows differences, for instance, between genders. Postmodern relativism puts the softer, social sciences on the same footing as hard science and thereby erodes the value of objective research.

We maintain that without believing in the objective validity of scientific reason, our society loses its way. The feel-good fallacies of the anti-scientific left have led a substantial portion of American society astray. They've brought on wrongheaded policies and positions on such crucial subjects as the food we eat; the environment we occupy; our policies with respect to energy, medicine, and business; and even the way our scientists operate. This must change. That is the argument of this book—and such a book must begin with the very president progressives elected to execute their preferred policies: Barack Obama.

Chapter Two

THE STATE OF THE "PRO-SCIENCE" UNION

How Does the Obama Administration Measure Up?

IN HIS 2009 INAUGURAL ADDRESS, President Barack Obama promised to "restore science to its rightful place." Millions rallied to his cause, frustrated by what they perceived as the George W. Bush administration's outright hostility toward science, particularly his policies on human embryonic stem cells, public health, environmental issues, and climate change.[1] We're not in the business of defending Republicans or conservatives—and so we leave those claims to stand on their own. But let's take a closer look at Obama's vow. The new president made a very serious promise about upholding science—going so far as to appoint a Nobel Prize–winning scientist to his cabinet. Four years later, how did he do?

Unfortunately, not well. Many of President Obama's major policy decisions have flown right in the face of responsible science. Being a political progressive does not necessarily make someone pro-science, any more than being a conservative automatically makes someone anti-government. After nearly a full term of a supposedly "pro-science" presidency, Barack Obama simply replaced George W. Bush's set of conservative anti-science policies with progressive ones.

Barack Obama vs. Embryonic Stem Cells

At the very top of President Obama's to-do list was to open the floodgates of human embryonic stem cell (hESC) research. During the Bush years, the funding of hESC research was often cited as one of the primary examples of how Republicans were pursuing anti-science policies.

How did this controversy begin? What even *is* hESC research, and why does it matter? To answer that question, we must look back to the contentious early days of President Bill Clinton's administration.

In 1995, Republicans were riding high. They not only controlled the Senate, but they also had just the year before taken a majority in the House of Representatives for the first time since the 1950s. Newt Gingrich was Speaker of the House, and under his leadership Republicans were attempting to steer policy in a new ideological direction. They tackled stem cell research with the tricky-sounding Dickey-Wicker Amendment. This unassuming rider, added to a separate appropriations bill by pro-life Representative Jay Dickey (R-AR), forbade the federal government from funding any research that involved the creation or destruction of embryos. The culture war over abortion may have been raging at full tilt, but the Dickey-Wicker Amendment had little effect, and science marched onward without problems.

Everything changed in November 1998. Researchers led by Dr. James Thomson of the University of Wisconsin reported that they had isolated human embryonic stem cells for the first time.[2] These hESCs were derived from several-day-old embryos. Because these cells retained the ability to differentiate into any cell type in the body, their enormous therapeutic potential was immediately recognized. Perhaps, someday, hESCs could be used to treat everything from paralysis to Parkinson's disease. However, because the hESCs were derived from human embryos, they were anathema to pro-life activists, as well as legally forbidden under the Dickey-Wicker amendment. But the Geron Corporation and the Wisconsin Alumni Research Foundation had funded Dr. Thomson's initial work privately, leaving the government's hands clean.

The controversy began in earnest when government-financed researchers expressed interest in examining the potential of hESCs. In 1999, the National Institutes of Health (NIH) sought a legal opinion on the matter. The NIH was told by the Department of Health and Human Services that public funds could be used to study hESCs under the Dickey-Wicker Amendment as long as the derivation of the cells, which required destruction of embryos, was carried out by the private sector. In August 2000, just before a new president was to be elected, the NIH released its final guidelines for derivation of the cells and solicited hESC research grants—complete with the clever workaround to dodge Dickey-Wicker.

Republicans were generally in favor of hESC research,[3] because they had always been in favor of stem cell research, and that attitude largely continues to this day, with 58 percent in support of it.[4] They had consistently supported stem cell research for the previous thirty years. George W. Bush, then a candidate for the presidency, disagreed about federally funding hESC research. He was opposed to this new technology on moral grounds. After winning the presidency, Bush acted like the "compassionate conservative" that he had campaigned as and tried to resolve the issue cautiously and deliberatively. Through much of 2001, he solicited ideas on how to compromise between the scientific and ethical issues. By then, the NIH guidelines had received more than 50,000 comments. Though some Republicans and religious groups were against it, staunch conservatives such as Senator Strom Thurmond (R-SC), Senator Connie Mack (R-FL), and Senator Orrin Hatch (R-UT) all favored hESC funding. In August 2001, the president followed the advice of Senator Bill Frist (R-TN), a surgeon, and split the difference.[5] Via executive order, Bush compromised with both camps and allowed federal funding for hESC research but only for pre-existing cell lines.[6] In other words, federal funds were available to conduct research but not to create new hESC lines. The acting director of the NIH, Ruth Kirschstein, MD, applauded his decision, writing, "We are pleased with the President's decision to allow the use of Federal funds for important basic research on human embryonic stem cells."[7]

Progressives did not like it. Activists and pundits have since often denounced the decision as a "ban" on hESC research, despite its not having been federally funded previously at all. Yet the claim of a ban isn't true: Bush's decision placed limits on hESC lines that were eligible for federal funding, but it did not ban anything. States and private industry were completely free to pursue hESC research and did, just as they had done originally in isolating hESCs.

Bush's solution was simple enough—but as with any issue where abortion is even tangentially involved, the controversy raged on. Personally opposed to revising his position, Bush twice vetoed the bipartisan Stem Cell Research Enhancement Act, which would have allowed the use of federal funds to create new hESC lines.[8] Meanwhile, states took matters into their own hands. California, for one, proceeded to allocate $3 billion for hESC research, including creation of new lines.

And then came the dawning of the Obama era. As 2008 ended, charged-up progressives were ecstatic. They finally had a president willing to stand up for stem cell research, and less than a month after his inauguration, President Obama gave them what they were asking for. On March 9, 2009, he signed an executive order that allowed federal funds to be used for the creation of new stem cell lines. But three years after Obama issued his order, hESC research was not much further along than it had been during the Bush administration. What went wrong?

Obama's 2009 executive order allowed federal funding for newly created stem cell lines. It instructed the NIH to create guidelines on the funding of hESC research just as it had done in 2000. Because of this decision, as of January 2012 there were 142 hESC lines in the NIH registry available for federal funding,[9] an increase over the 21 hESC lines eligible in 2001. However, continued NIH restrictions (e.g., requiring donor consent) made the number of hESC lines available for federal funds far less than the approximately 760 lines existing globally.[10] Obama's executive order was no different from any other political action—it looked good and it pleased the base, but it didn't do much of anything to advance the needs of the scientists involved.

That's not all: hESCs are just the best known of the research techniques that have become politicized. Cutting-edge biological science has created brand new ethical dilemmas. A process known as somatic cell nuclear transfer (SCNT)—used in what is called "therapeutic cloning"—is still not eligible for federal funding.[11] This process was famously used to clone Dolly the sheep. The scientific community supports this technique as an important new tool for studying genetic disease or even, someday, developing the ability to regenerate entire organs. To perform SCNT, a nucleus is extracted from a body cell and transplanted into an egg cell whose nucleus has been removed. The purpose of the procedure is to create hESCs that are then used to generate tissue for scientific analysis or regenerative medicine. But this isn't quite the dystopian science fiction scenario it may seem: the hESCs are *not* used to clone new human beings.

The scientific community would love for federal funding to be available for this particular research technique. But in spite of

Obama's symbolic move to decrease restrictions on high-profile human embryonic stem cell research, the NIH under the Obama administration still does not fund actual cutting-edge biological research. And there is little reason to believe that any moves are being made to accept the scientific consensus.

Advocates of Obama's policies love to portray them as a revolutionary departure from those of the ideological Bush administration, but in reality the new policy is barely an incremental improvement. Bush was harshly criticized by the left for being "anti-science" because his policies were not in line with those favored by the scientific community—the same as Obama's unwillingness to fund SCNT. It's strange that progressives refuse to consider Obama just as anti-science.

Barack Obama vs. Vaccines

In 2008, Obama ventured into the realm of vaccines and autism. While on the campaign trail, he said,

> We've seen just a skyrocketing autism rate. Some people are suspicious that it's connected to the vaccines. This person included. The science right now is inconclusive, but we have to research it.[12]

Wrong. The science was decidedly *not* inconclusive as of 2008. Discredited researcher Andrew Wakefield's original twelve-person "study" linking vaccines to autism was published in 1998, but by 2002 the medical community had thoroughly debunked Wakefield's claim. That year, the *New England Journal of Medicine* published an enormous epidemiological study, including more than 537,000 people, that demonstrated no link between vaccines and autism.[13] In reality, the notion of there ever having been a "controversy" in regard to vaccination is fallacious; the medical and scientific communities have always endorsed vaccines as one of the basic foundations of public health.

By the time Barack Obama was running for president, the vaccine-autism link had been comprehensively dismissed (and its underlying research was eventually found to be fraudulent as well). So why did Obama claim the science was unclear? Perhaps he wasn't up-to-date on the latest findings on the subject.* For someone with such enormous influence over public health policy to be ignorant of basic medical facts is frightening indeed.

Even if we give Obama the benefit of the doubt on his autism gaffe, he had a troubling relationship with vaccines even after assuming the presidency. In 2009, when the world was bracing for the H1N1 swine flu pandemic, the United States experienced a vaccine shortage. Former deputy commissioner of the Food and Drug Administration (FDA) Dr. Scott Gottlieb pinned the blame on a sluggish and overly cautious FDA for holding up production, but the Obama administration ultimately bears responsibility because the FDA is under the purview of the Department of Health and Human Services.[14]

In the midst of the swine flu panic, two decisions in particular flew in the face of immunology and medical science. The first was the refusal to allow adjuvants—which boost a person's immune response—to be mixed with the vaccine. Using adjuvants decreases the amount of raw material required per shot, and that may have quadrupled the U.S. vaccine supply.[15] Additionally, adjuvants are used in Canada and some European nations and are endorsed by the World Health Organization as an effective method to extend the vaccine supply.[16] The second bad decision was the insistence on producing single-dose, as opposed to multi-dose, vials. The

---

*Senator John McCain was also afflicted with this same anti-science position during the presidential race in 2008.

reason given for this decision was that single-dose vials would contain less thimerosal—the preservative that opponents of vaccines believed causes autism. It does not. On the campaign trail, candidate Obama had ignored basic medical facts and pandered to the anti-vaccine crowd. In office, his administration had failed to be either sensible or scientific when it came to vaccinations. As a result, production chains backed up and many Americans couldn't get flu shots.[17]

By April 2010, the Centers for Disease Control and Prevention (CDC) estimated that 61 million Americans had been infected with H1N1 swine flu, 274,000 had been hospitalized, and 12,470 had died.[18] It's not possible to calculate how many of those illnesses and deaths could have been prevented if the Obama administration's pandering to the irrational concerns of progressives hadn't contributed to the vaccine shortage. However, what *is* certain is that the administration has been far more anti-science than pundits and progressive activists would like to believe.

Obama and the BP Oil Spill

In April 2010, the *Deepwater Horizon* oil-drilling platform in the Gulf of Mexico operated by BP exploded, causing one of the worst oil spills in world history. For presidential leadership to be effective during such a disaster, relevant information must be accurately and transparently produced—and it must be delivered to scientists as quickly and reliably as possible. How did the Obama administration do?

Before we answer that, let's remember that President George W. Bush was constantly criticized for manipulating scientific data for political and ideological purposes.[19] This was such a big issue for President Obama that he even addressed it during the signing ceremony for his executive order regarding stem cells:

[Promoting science] is about letting scientists like those here today do their jobs, free from manipulation or coercion, and listening to what they tell us, even when it's inconvenient—especially when it's inconvenient. It is about ensuring that scientific data is never distorted or concealed to serve a political agenda—and that we make scientific decisions based on facts, not ideology.[20]

We agree with these standards. Did Obama live up to them during one of the biggest crises of his administration, the BP oil spill?

No. In fact, the Obama administration did exactly what they accused the Bush administration of doing: they manipulated data and withheld information from scientists. A devastating article published in the *Los Angeles Times* described independent government reports that were issued following the spill: taken together, two of the reports painted a picture of a U.S. government that was unprepared to deal with a catastrophic spill like BP. And the portrait of an administration that withheld information from the public and, more specifically, scientists, about how much oil was getting into the water, how much remained, and how such estimates were calculated appeared to contradict Obama's pledge to make government more transparent and trustworthy.[21]

It gets worse. A panel of outside experts reviewed drilling safety recommendations put together by the Department of the Interior. After the scientists approved the draft, the White House altered the contents of the document to include a moratorium on offshore oil drilling—thereby giving the false impression that the panel also approved of the moratorium.[22]

Essentially, when things got tough, the Obama administration did what many other politicians and bureaucrats do: withhold and manipulate information. Obama did everything *except* "restore science to its rightful place."

Obama's FDA vs. Science

Stem cells, cloning, vaccines, and the BP oil spill are just the be-ginning of the Obama administration's inconsistent record on sci-entific issues. Another puzzling public health decision under the administration is the FDA's stubborn opposition to electronic cig-arettes. "E-cigarettes," as they are known, first entered circulation in the past decade. They work by heating and vaporizing a liquid solution containing nicotine. The "smoke" that emerges contains only nicotine—the same addictive chemical in smoking cessation gums and patches—and none of the carcinogens found in to-bacco smoke. E-cigarettes could be a valuable tool in helping hard-core smokers reduce or eliminate their consumption of reg-ular cigarettes.[23] The result would be a shrinking of the number of smokers and, consequently, of the incidence of lung cancer. A win-win situation.

Yet that outcome is not to be had under the Obama adminis-tration. These days, the same progressives who correctly rail against the absurdity of abstinence-only sex education bizarrely endorse abstinence-only smoking education. Instead of embrac-ing e-cigarettes, Obama's FDA has put up roadblocks without providing scientific evidence in support of its opposition.[24] Critics have bluntly suggested that the FDA's policy is killing smokers who want to quit.[25] Any way you look at it, real science has noth-ing to do with the policy the FDA has chosen. The policy is in defiance of science.

Nor is it the only one with troubling implications. The routine usage of antibiotics in animal feed has become an extremely im-portant public health issue. Farmers regularly provide antibiotics to their animals in order to prevent infections and to fatten them up for slaughter. It's a dangerous practice. These additives pro-mote the growth of antibiotic-resistant bacteria. When these super-bacteria are passed on to humans, treatment may be difficult if not impossible. The largely human-made problem of antibiotic-resistant "super-bugs" is one of the most active areas in modern microbiology research, as it has enormous implications for public safety. Yet the problem could be reduced with a simple, straightforward policy move: ban antibiotics for animals unless they are being treated for an active infection.

It's not as though the FDA doesn't know that this is an option. Under the Bush administration, the FDA wanted to limit a particular class of antibiotics called cephalosporins, which are used in animals prophylactically. But Bush overruled the FDA after farmers and drug companies complained.[26] The Obama administration reversed this action, and the FDA got its wish. But just as in the case of Obama's stem cell executive order, this victory actually meant very little. Cephalosporins aren't even in common use. They constitute less than 0.2 percent of all antibiotics used by farmers.[27] When it came to the remaining 99.8 percent, Obama's FDA had no appetite for striking any further blows for good science and policy. Representative Louise Slaughter (D-NY), the only microbiologist in Congress, remarked, "This is a modest first step by the FDA, but we're really just looking at the tip of the iceberg. We don't have time for the FDA to ploddingly take half-measures."[28]

Yet a half measure is what we got. For all practical purposes, an anti-science decision by the Bush administration was largely kept in place by an anti-science Obama administration.

Barack Obama vs. Nuclear Waste

One of the most nakedly political decisions of the Obama administration was the closure of Yucca Mountain, a facility in Nevada designed to hold the nation's nuclear waste. After great expense and extensive political wrangling, the facility had been built and was essentially ready to go. In a single, politically calculated move,

Obama upended nearly twenty-five years of research and wasted $13.5 billion of taxpayer money.[29] Why? Because his political ally, Senator Harry Reid of Nevada, didn't want the nuclear waste in his state—and he was facing a tough reelection in 2010. So President Obama canceled the project and claimed science was on his side. Far from it. At work here was the evil twin of politicized science: the "scientization" of politics.

Yucca Mountain had been a political football since its inception in 1987. The need for a place to store dangerous nuclear waste was as clear then as it is now. The science on this issue is straightforward: placing all nuclear waste in a single location, deep underground, in the unpopulated middle of nowhere is far safer than keeping it on-site at nuclear power plants in spent fuel pools. This fact was brought to the forefront during the meltdown of the Fukushima Dai-ichi nuclear power plant following the tsunami that hit Japan in March 2011. Scientists feared the worst: that the pools used to cool the fuel rods would stop working and there would be an even larger release of radioactive waste. Thankfully, the fuel pools survived the crisis more or less intact. Even so, consider that spent fuel pools in the United States have even more radioactive material than the ones at Fukushima[30]—a recipe for disaster if another natural catastrophe occurs.

In the late 1980s, three places were selected as candidates for a national disposal site. The Texas and Washington State sites were eliminated because powerful politicians from those states—House Speaker Jim Wright (D-TX) and House Majority Leader Tom Foley (D-WA)—imposed their political will. President Ronald Reagan also nixed a potential site to be located in the eastern United States, fearful of the pushback from voters.[31] That's how Nevada's Yucca Mountain got stuck with the waste. During this two-decade-long debacle, science played only a minor role; politics mainly drove the process.

Even so, Yucca Mountain is probably the best option currently available. It is isolated, it is relatively impregnable thanks to its geography, and it has been extensively studied for years. Not only that, by the time Obama was elected, the facility was already built. For all practical purposes, it was and still is the country's only viable option for long-term waste disposal. Yet from the moment he took office, Obama simply did not consider these facts. Instead, he fully intended to shut down the project from Day One.

Gregory Jaczko, originally appointed by George W. Bush to the Nuclear Regulatory Commission (NRC)—which oversees the nuclear power industry—was elevated to chairman by President Obama in May 2009. The NRC supposedly regulates the nuclear industry without bias or partisanship, but Jaczko, a former aide to anti–Yucca Mountain Senator Harry Reid, did everything in his power to shut down the project.[32] That even included withholding information from his colleagues on the commission.[33]

Obama also appointed a "blue ribbon commission" to study the problem of nuclear waste, but then, according to the *Bulletin of the Atomic Scientists*, commission members were specifically instructed *not* to issue a ruling on the suitability of Yucca Mountain:

> The commission asked the Energy Department to provide its justification for shutting down Yucca Mountain, and science explicitly was not part of Energy's reasoning. Instead, the Energy Department claimed that because of Nevada's objections to the site, the controversy would never be resolved.[34]

This is particularly disappointing because Obama's secretary of energy is Steven Chu, a Nobel Prize–winning physicist. Chu justified the closing of Yucca Mountain by citing the safety of on-site dry cask storage and the possibility of using fast-neutron reactors to consume the longer-lived radioactive waste.[35] But this sounds like a desperate attempt to find a scientific rationalization for what was always a political decision, the very definition of the scientization of politics. Given the golden opportunity of following through on his promise of restoring science to its rightful place, President Obama once again chose the opposite, closing the Yucca Mountain site as a political gift to Harry Reid. Now, the rest of the nation must deal with the consequences.

There are 104 operating nuclear power plants in the United States. Instead of there being one secure location to place the waste, it is scattered throughout the country. Despite inevitable flaws, storing nuclear waste in Yucca Mountain is a vastly safer option than the current disorganized policies. Those who fear radioactive leakage from Yucca Mountain are bizarrely untroubled by the much greater risk of leaving the waste where it is. And since we can't wish the nuclear waste away, we must come up with a viable long-term solution. It took scientists twenty-five years to overcome the vagaries of politics to concoct the solution of Yucca Mountain, and it took less than four months for the Obama administration to scuttle their hard work.[36] It may be another twenty-five years before someone comes up with a new idea.

Barack Obama vs. the Environment

Rhetoric and campaign promises aside, President Obama's record on the environment has been as inconsistent and disappointing as his record has been on any of the other scientific issues we've already raised. The much-ballyhooed "Cash for Clunkers" program was an initiative in which the government tried to stimulate the economy by subsidizing, through rebates, the purchase of newer and supposedly greener vehicles. The government paid for people to take cars that were getting approximately 15.8 miles per gallon (mpg) and trade them in for cars that averaged around

25 mpg. With 690,000 cars traded in, this translates to saving an aggregate 12,000 barrels of oil per day.[37] In a country that uses 9 million barrels of oil on any ordinary day, this is at best a drop in the bucket.[38] Worse, a separate analysis revealed that newly purchased vehicles were only 0.23 mpg more efficient than if the program had not existed, meaning that the benefit in terms of reducing fossil fuel consumption was marginal at best.[39] Also, the engines of the older cars were destroyed rather than reused. But the formidable environmental cost of manufacturing brand new cars did not factor into anyone's calculation.

You might wonder if the trade-off was worth it. Even though the program wasn't as green as the Obama administration said it was, did Cash for Clunkers at least stimulate the economy? Not so much. People who received rebates were already planning to purchase new cars. Instead of generating new demand for vehicles—which truly would have been a "stimulus"—the program concentrated existing demand into the lifetime of the government program.[40] Auto dealerships and car manufacturers got a burst of sales, but not substantially more than they would have had over a longer period of time. The short-term cost to taxpayers, according to an analysis by automotive website Edmunds.com, was $24,000 for each clunker.[41] The long-term benefit was questionable at best and, at worst, nil.

Of course, at the time "clunkernomics" was extremely popular with the American people. The government was handing out billions of dollars in "free" money, and other taxpayers were stuck with the bill, so what was not to like? It just wasn't good economic or scientific policy. With a piece of legislation specifically intended to both help green the planet and boost the economy, the Obama administration actually did neither. It was an impressive feat of dysfunction. It's not often that politicians get the opportunity to violate two entirely different fields, science and economics, while claiming to advocate for both.

Yet time and again, in the years that followed, Obama let his political instincts lead him into anti-science environmental stances. After the EPA recommended that the United States tighten its smog standards, in September 2011 Obama rejected the proposal. In a move that resembled the worst anti-scientific tendencies of George Bush, Obama overruled the recommendations of his own team of scientists, pronouncing business and economic concerns more important than clean air.[42] Environmentalists routinely criticized Bush for placing corporate interests above the recommendations of scientists. Then Obama threw up his hands and did exactly the same thing. Hard-core environmentalists groused, but few journalists stepped up to decry this all-too-familiar intrusion of politics into the world of science.

This isn't to say that President Obama's record on the environment is entirely composed of failures. He is to be commended for ending the ridiculous subsidization of corn ethanol. Posited as a possible replacement for fossil fuels, it is in fact far inferior and, if anything, *less* green. In an article appropriately titled "Biofools," *The Economist* explained that the process of converting crops to biofuel releases nitrous oxide (N_2O), a greenhouse gas with about three hundred times the global warming potential of carbon dioxide (CO_2).[43]

But though politics is rarely scientific, all politics *is* local—and the corn for ethanol is grown in massive quantities by Iowa farmers. No wonder that Obama supported the subsidy during his campaigning for the 2008 election. While his team maintained extensive ties with the ethanol industry,[44] he benefited from the votes of Iowa corn farmers who were eager for a federal subsidy, and he ultimately won the Democratic Iowa caucuses with their support. He's hardly the first politician to kneel at the altar of the Iowa farmer, offering worship and praise for corn ethanol subsidies. As recently as 2011, Republican presidential candidate Mitt Romney also voiced support for this program.[45] Even Al

Gore admitted endorsing biofuels for years simply to get votes.[46] Thankfully, science won a very rare victory over politics in December 2011 when Obama reversed course and signed a bill from Congress (with a Republican-controlled House, no less) that pointedly did not renew the $6-billion-per-year, corn-fueled absurdity.[47] Nevertheless, politics reigned supreme when it came to one of the most consequential scientific crises of the modern era: the prospect of a changing climate. Most likely everyone reading this book remembers when Barack Obama said on the campaign trail, "This was the moment when the rise of the oceans began to slow and our planet began to heal."[48] We have all heard some questionable statements by politicians that later came back to haunt them. ("Read my lips: no new taxes" and "I did not have sexual relations with that woman" are among two famous recent examples.) But Obama's high-soaring rhetoric really raised the bar as far as environmental expectations were concerned. How did President Obama do? Are his climate policies reversing sea level rise and causing the planet to heal?

No. On the issue of climate change, President Obama did almost nothing. Despite his campaign promises and the top priority Democrats have claimed to place on climate change for the past several years, there has been very little progress to speak of. Climate change legislation is often stymied by furious opposition from businesses and both Republicans and Democrats who would prefer that Congress—and not unelected bureaucrats in the EPA—make laws. To bypass this, in March 2012 the EPA placed strict limits on carbon emissions from new coal power plants.[49] This is potentially a good first step toward turning the tide, but it hardly fulfills his lofty promise to reverse the rising sea level.

Republican opposition often gets the blame for inaction on climate change. Yet following the 2008 election, the Democrats held overwhelming majorities in the House of Representatives and in the Senate. For a time, they even held a rare filibuster-proof, 60-seat majority. A cap-and-trade bill, claiming to reduce greenhouse gas emissions by 17 percent by 2020 and promote green jobs,[50] passed the House, but it never had a chance in the Senate. Even some Democrats were against it. As a candidate, Democratic Senator Joe Manchin, who represents the coal-mining state of West Virginia, famously ran a commercial in which he shot the cap-and-trade bill with a rifle.[51] So the Obama administration made the strategic decision to let it die. Yet the blame nevertheless fell almost entirely on the right. Much as environmentalists and progressives in the media bashed Bush and the Republicans for not ratifying the climate treaty known as the Kyoto Protocol, the silence was deafening when Obama failed on this issue as well.

Given the high expectations that President Obama set for himself when he entered the Oval Office, it is clear that he has not even come close to living up to them. Instead of promising to restore science to its rightful place, he should have promised to "continue the political assault on good science, but this time from a left-wing viewpoint." That would have been a promise he could easily fulfill.

Chapter Three

ORGANIC FOOD: THE HOLY EUCHARIST OF ENVIRONMENTALISM

Don't Buy That $4 Banana

THESE DAYS, WHOLE FOODS seems to be the height of fashion. Its selection of products is impressive and exotic. The stores are enormous and well stocked. And best of all, the food is as good for you as it is for the earth. So what if it costs just a little bit more? You simply can't put a price tag on doing right by the environment. Repeat after me, "*Shopping at Whole Foods will make you healthier, smarter, and more moral than everyone else.*"

At least that is what the Whole Foods crowd believes. Imagine the ideal customers: well educated, trim, upper class, armed with reusable bags and organic soy lattes from the local nonprofit café. They walk out of their Priuses and into their local Whole Foods stores with an attitude of reverence, as if this were a utopian vision of a progressive future—a church where they can pay tribute to all of their deepest held ideas about food, health, and the environment.

The reality is Whole Foods customers may be more well-off and more likely to hold higher degrees than the average shopper—but what really unites them is that they are just more gullible than everybody else. At a Whole Foods market in Seattle,

customers are greeted by signs that warn them to avoid genetically modified (GM) food. These warnings conveniently fail to
mention that there is not even a single documented case of GM
food causing a stomachache, let alone any lingering health problems. Whole Foods promotes organic food as a wholesome alternative to the poison customers would buy at one of their
conventional grocery store competitors. It's a convenient way to
make lifelong, dedicated customers out of its patrons. This business model is founded upon selling a feel-good, utopian lifestyle
based on a lie. And that lie is the emotionally persuasive, but unscientific, idea that natural things are good for you (and, by extension, unnatural things are bad for you).

It is not a surprise, then, that public health officials are starting to recognize a pattern. Pull out a map of the United States,
and stick in a pin wherever there is a Whole Foods. That community will also likely have a strong anti-vaccine movement.[1]
Perhaps even less surprising, given the sort of progressive who
likes to shop at Whole Foods, is the fact that 81 percent of counties that have a Whole Foods voted for Barack Obama, compared to just 36 percent of counties with a Cracker Barrel.[2] Of
course, Whole Foods will claim that it picks store locations based
on income level. But whether the company intends to or not, the
reality is it places stores in cities that have a large population of
progressives.

How can such educated, rich people be so uninformed about
food? Ironically, Whole Foods founder John Mackey is an Ayn
Rand fan who practices yoga.[3] We could call him a "granola conservative." He created an enormously successful food store chain
by appealing to those people who are health and environmentally
conscious, even to the point of rejecting technology in favor of
what is "natural." This simpleminded reasoning, often endorsed
by progressive thinkers, largely forms the basis of the organic
food movement as well as the anti-vaccine movement. Even

though the worship of all things natural sounds like a nice philosophy, science has something else to say about it.

The War on Genetically Modified Organisms

The vituperative public outcry against genetically modified organisms (GMOs) is reminiscent of how the first Neanderthal must have reacted to fire. (Although we shouldn't be too hard on Neanderthals—the evidence seems to indicate they eventually accepted science and used fire regularly, making them smarter about physics than anti-science progressives are about genetics.) The opposition to GMOs is rooted both in the "natural is better" myth and in ignorance of basic science. Ignorance breeds fear, and fear breeds opposition. With better science literacy, in twenty years if we are lucky, society will be able to look back and wonder what all the fuss was about.

The science behind GMOs is straightforward: find a gene that is useful, and insert it into an unsuspecting organism we care about. Before this technology was refined by Herbert Boyer and Stanley Cohen, when they created the first successful recombinant DNA organism, plants and animals had long been "genetically modified" by breeding, but it was a time-consuming, inefficient process. Farmers identified advantageous traits in their crops and livestock and then used selective breeding to produce more robust offspring. This took generations and generations to work.

Another method of historic genetic modification was less targeted: scientists have attempted to randomly generate mutations, for example in plants, in the hope of creating a new beneficial trait. In the 1950s and 1960s, scientists blasted plants with radiation, crossed their fingers, and hoped for something awesome. Lucky for them, they generated a fungus-resistant peppermint plant that is still used to this day to make chewing gum and toothpaste.[4]

Both methods have enormous drawbacks. First, selective breeding is time-consuming and non-specific. If mommy goat has a good gene and daddy goat has another good gene, selective breeding does not guarantee that the kids will get both genes. Generating random mutations is just as non-specific. It is somewhat like smashing DNA with a sledgehammer and hoping something interesting occurs. This kind of random mutagenesis may sometimes yield interesting results, but it is practical only for plants—blasting little Billy the goat with gamma rays remains an ethically questionable tactic. Finally, and most importantly, it is not possible using these methods to create entirely new metabolic properties. For example, if a farmer wants a goat that glows green so that he can find the animal at nighttime, no amount of selective breeding or random mutagenesis will accomplish that. Genetic modification solves all of these problems.

One of the classic examples of genetic modification involves a gene encoding Bt toxin. This comes from the bacterium *Bacillus thuringiensis* (hence the name "Bt"). The toxin carried by this bacterium has a very interesting property: it can kill caterpillars but is completely harmless to humans. In the 1990s, scientists cloned this gene into commercial crops, and—voilà!—plants naturally resistant to caterpillars were born. This innovation lowered the demand for insecticides, which pollute the environment and are harmful to people. In 1998, the U.S. cotton crop alone required 450,000 kilograms (kg) less insecticide than would have been required had the crops been grown without genetic modification.[5]

It gets better. Superior varieties of drought-resistant wheat are being developed to help feed the planet's growing population and to combat the effects of global climate change.[6] "Golden rice," supplemented with vitamin A, is being developed to combat vitamin A deficiency, a disease that leads to blindness and contributes to hundreds of thousands of deaths worldwide.[7] Wilt-resistant bananas, which are able to withstand assault from the bacterium *Xanthomonas campestris*, are being tested in Uganda with the hope they will save the country's staple food crop.[8] Indeed, GM crops are spreading across the entire continent of Africa because of the numerous advantages they confer over conventional crops. Scientists have even proposed creating "greener" trees that are better at sequestering carbon than normal trees, helping the fight against global warming.[9] Excited? We haven't even mentioned the GM animals yet.

In order to understand everything from basic genetics to immunology, research scientists utilize thousands of different strains (that is, "breeds") of GM animals. Deleting and inserting genes in mice, fruit flies, and worms are the most frequent manipulations. Scientists genetically modify those "model" organisms in order to examine the basic biology of living things. Additionally, the technique allows the simulation of human disease in the lab. Experiments using genetically modified organisms are now so commonplace that without them, biological and medical science would cease to function. We can thank genetic modification for illuminating difficult concepts such as circadian rhythm, cancer, embryonic development, and bacterial pathogenesis. But don't think that only scientists benefit from GM animals. Malaria-resistant mosquitoes[10] and chickens incapable of spreading influenza[11] have the potential to save millions of lives.

Genetic modification therefore represents a veritable revolution in agriculture, biology, and public health. GMOs can serve as a valuable research tool, help combat disease, reduce the use of harmful pesticides, and increase agricultural efficiency, which is dearly needed if we are to feed a planet whose population is expected to hit 9 billion by 2050. Besides, GMOs are already everywhere. They are so prevalent in the United States, for instance, that around 70 to 75 percent of foods at the grocery store contain at least one genetically modified ingredient.[12]

So if GMOs are really as great as we say they are, then why are so many progressives opposed to them? As we already mentioned, fear bred from scientific ignorance is likely the biggest reason. But there are other reasons to oppose GMOs. Are these reasons legitimate? No, but at least they are more intellectual than the sheer, unsupported resistance to science you'll hear voiced (and see misspelled on signs) at most anti-GMO rallies. Let's take a look at, and then debunk, the most common anti-GMO arguments made.

First, opponents like to start by launching a broadside against Big Business. They claim that biotech industries are profiting off poor farmers by forcing them to buy patented seeds year after year. It is true that patent-holders like to make money, but they cannot *force* farmers to adopt their technology. Farmers can continue using their own methods, or they can switch to a more advanced technology, which is more expensive. That's how the marketplace works. However, farming—just like economics—is not a zero-sum game. GM crops may require a bit more in setup costs, but the end result is increased profit for both the biotech firm and the farmer. This is exactly what surveys have shown. Small farmers in developing countries have benefited the most from GMOs because, in spite of increased initial expenditures, they've seen a lower overall cost and dramatically increased yields.[13] Though seeds are more expensive, farmers save money by using less pesticide, another net benefit for the environment. Developed countries do not see dramatically higher crop yields (probably because their agricultural technology is already advanced), but they likely reap long-term environmental benefits in the form of improved soil and water quality.[14] So much for the economic argument.

Second, opponents claim that GMOs are not natural and hence are an affront to Mother Nature. But everything humanity does is in defiance of Mother Nature. Agriculture itself, including selective breeding for thousands of years, is the exact same kind of manipulation of nature. Does anyone really think a labradoodle is natural? And what we know today as corn is completely unnatural since it took millennia of careful selective breeding to produce it. If you enjoy tomatoes bigger than the size of your thumb, thank genetic manipulation. The only truly "natural" way of living is to be a hunter-gatherer. Most people probably aren't ready for this lifestyle adjustment.

Third, naysayers believe that GMOs pose a threat to human health and the environment—except scientists have examined this possibility for decades and found no basis for concern.

Activists make the claim that GMOs are not sufficiently tested to prove they are safe for human consumption or for the environment. This argument is a red herring. The FDA uses a policy known as "substantial equivalence" to determine the safety of new foods. Basically, the policy asks one question: Is GM food nutritionally the same as the unmodified, natural version?[15] If it is, the GM food is considered safe and not subject to further scrutiny. If the GM food is different, it might need FDA approval, depending on what exactly is being added to the food.[16] If a company wants to use strawberry genes to make salmon fruity-flavored, it may not require FDA approval because strawberries are already considered safe. But if the company wants to make the salmon glow in the dark, and therefore easier to catch, then it has to go the extra mile and prove to the FDA that the glow-in-the-dark protein is not toxic. Will this glow in the dark gene trigger an allergic response? Will it cause cancer? Will people start to glow in the dark if they eat it? All of these questions must be answered for this genetic modification to receive FDA approval. The regulatory burden increases if the company then wants to stock ponds and rivers with this fish—it then has to answer to the U.S. Department of Agriculture (USDA) and the Environmental Protection Agency (EPA), both of which oversee the use of

GMOs in nature. In short, biotech companies must navigate several bureaucratic obstacles before their products are allowed to go to market.

Of course, GMOs should be closely monitored or regulated. Long-term studies should continue to be conducted to confirm that there is no harm to human health or the environment. Like many scientific advances, issues of risk must be weighed against rewards. But contrary to the claims of opponents of genetic advances, there's no evidence after years of testing that GMOs pose any risk. There is simply no evidence to suggest we should stop using one of our best hopes to expand the global food supply, become more agriculturally efficient, and reduce carbon emissions. Based on the current evidence, it is crystal clear that the demonstrated benefits of GMOs far outweigh the hypothetical costs. They should be fully utilized until society has a good reason not to use them. Any other conclusion is based on speculation and is unscientific.

And just in case you're wondering, there *are* GM fish that glow in the dark.[17]

Organic Food: The Holy Eucharist of Environmentalism

Michael Crichton, famed author of *Jurassic Park*, was not a big fan of environmentalism. In a 2003 speech, he summarized the religious nature of today's environmental movement:

> We are energy sinners, doomed to die, unless we seek salvation, which is now called sustainability. Sustainability is salvation in the church of the environment, just as organic food is its communion, that pesticide-free wafer that the right people with the right beliefs imbibe.[18]

It is probably no coincidence that Crichton went on to write a novel about mass-murdering eco-terrorists.

Those who support organic food state that it is healthier and tastes better. The latter claim is not really important from a public health standpoint, but that did not prevent Penn & Teller from examining the claim in their Showtime series, whose title is left to an endnote but, as a hint, is synonymous with "bovine excrement."[19] Theirs was probably not the most scientifically rigorous analysis, but it made some good points. Using blind taste tests, they found that the average organic food shopper could not tell the difference between organic and conventional food. Many of them even preferred the taste of conventionally grown food. The most amusing segment asked organic shoppers to taste two "different" bananas. The first banana was conventional and the second organic. Or so they were told. In actuality, it was the same conventionally grown banana cut in half. Predictably, nearly all the organic shoppers preferred the "organic" banana, demonstrating that the preference for organic food is largely driven by marketing and is therefore psychological. A more formal analysis reached a similar conclusion in regard to strawberries, demonstrating that taste testers could not discern a difference between organic and conventional varieties.[20] They *could*, however, discern an objective difference in cost, as the organic strawberry sold for two to three times the price of its conventional counterpart.

So what about the health claims? Is organic food actually healthier than conventional food? Nutritionally, conventional crops are likely to be just as healthy as organic crops. According to a study conducted by the University of Copenhagen, levels of antioxidants were identical between organic and conventional vegetables.[21] There might be a negligible health benefit from consuming grass-fed milk and beef, as both possess slightly lower saturated fat content.[22] Yet if healthy food is your primary concern, you should eat fish instead because that is by far the best source of omega-3 fatty acids and has fewer drawbacks than beef. Whether or not being "grass fed" justifies the price and the supposed moral superiority that comes with eating free-range cows allowed to wander around before their systematic execution is a cultural debate, not a scientific one. Regardless of how well the cow was treated beforehand, at the end of the day it still turns into a Happy Meal.

More worrisome for organic food proponents should be that both organic and conventional produce is sprayed. Yes, contrary to popular belief, organic farms can and do use pesticides. In fact, they often use Bt toxin as a spray (while hypocritically objecting to its use in GMOs). The only distinction is in the kind of pesticides. Organic farmers do not use *synthetic* pesticides. In their mind, this allows advocates for organic food to bemoan conventional produce as being coated in poisonous chemicals. Even so, the amount of synthetic pesticides found on conventional crops is far below what is considered safe for human consumption. Thus, organic crops offer no real health benefit in this regard.[23]

It bears mentioning that the pesticides declared safe for use on organic crops are also not any more environmentally friendly than synthetic pesticides are. Because these pesticides are often less effective, larger doses are required—and the more farmers use, the more unintended effects there are on the surrounding habitat.[24] No one on either side of this debate claims that any pesticide is completely safe for humans and the environment. Research indicates that excessive levels of some pesticides may act as endocrine disruptors, for instance. But the blanket claim that organic food is healthier because it is grown with fewer pesticides seems to be mostly hype. Unless you personally know the farmer who grew the food and inspected his agricultural practices, you are much better off just buying the cheapest produce at the local grocery store. Thus, pesticide use is just one of many dubious claims that the organic food movement makes about how much better its techniques are for the environment.

Some evidence does suggest organic farming can promote healthier soil and ecosystems,[25] but the downside is that farms' efficiency markedly decreases in terms of both crop yield and cost.[26] Because organic farmers stay away from conventional fertilizers, synthetic pesticides, and genetically modified organisms, the easiest way to increase the supply of organic food is to convert more acreage into farmland. More farming requires more space, so this is hardly better for the environment in the long run. As a side note, a higher cost of production combined with a decreased supply of food may help explain why candidate Obama was famously dismayed over the price of arugula at Whole Foods in 2008. Put simply, inefficiency, not superior product, is the reason Whole Foods is more expensive.

When other environmental effects are factored in, such as greenhouse gas emissions, organic farming becomes even less attractive. Organic wine, for instance, requires eighty times more fertilizer than conventional wine.[27] Impressive piles of manure have to be trucked in from somewhere; all this extra transport contributes to global warming. A study by the University of Alberta demonstrated that shipping food over great distances cancels any positive environmental effect from growing it organically.[28]

Perhaps the most stunning revelation about organic food is that just because something is labeled "organic" in the grocery store does not mean it actually *is* organic. Mischa Popoff, an organic food inspector, remains concerned that the USDA's National Organic Program (NOP) has failed to shore up federal oversight of organic food. As of early 2012, there was still no surprise field-testing in the organic food industry. Usually, all an organic farmer has to do in order to be certified organic is fill out paperwork.[29] Therefore, it should not come as a surprise that the Consumers Union once found that 25 percent of organic food in the United States was contaminated with traces of pesticides that were not allowed by any organic certification program.[30] Imported organic food is even worse. The United States continues to import allegedly organic food from China, which has a dismal food safety record. Chinese "organic" farmers commonly use pesticides and can purchase organic certification through illegal means.[31] Supposedly, an American firm called the OCIA (Organic Crop Improvement Association) was inspecting Chinese organic farmers. However, it was revealed that the OCIA, which was accredited by the USDA, was actually using Chinese government officials to inspect state-controlled farms.[32] To make matters worse, the OCIA follows the industry norm of collecting a small royalty on each organic sale it oversees, which means the association profits from the very farmers it is supposed to be regulating.[33] To the USDA's credit, it cut ties with the organization due to the blatant conflict of interest, but the OCIA remains one of North America's largest USDA-accredited organic certifiers. With this in mind, you might wonder just how much purportedly "organic" food has been sold that falsely bears a stamp of approval from the U.S. government.

Remarkably, despite the honesty problem that plagues the industry, organic activists still have the audacity to smear conventionally grown food from "Big Ag." For example, the Environmental Working Group published the "Dirty Dozen," a list of conventional foods it claims are contaminated with pesticides.[34] But science has something else to say about the matter. A paper in the *Journal of Toxicology* reported,

> It is concluded that (1) exposures to the most commonly detected pesticides on the twelve commodities pose negligible risks to consumers, (2) substitution of organic forms of the twelve commodities for conventional forms does not result in any appreciable reduction of consumer risks, and (3) the methodology used

by the environmental advocacy group to rank commodities with respect to pesticide risks lacks scientific credibility.[35]

Ouch. Unfortunately, this assessment serves as a nice summary for the entire organic food movement: it lacks scientific credibility.

If you listen to what organic advocates say, you would think that the food was grown, handpicked, and delivered to the local non-profit supermarket by the gods themselves. In reality, organic farmers have simply rejected the science that has made modern agriculture successful over the past several decades. What is the final verdict on organic food? Because more research needs to be done, a gracious answer would be that the jury is still deliberating. However, the foreman is ready to report on a decision, and it will not be in favor of organic farming.

Cows and the Café Confrontation

One day, co-author Alex was in a Seattle coffee shop that proudly and prominently displayed the fact that it used only rBST-free milk. Apparently, it might be in good company: Safeway's "Lucerne" dairy brand proudly denounces rBST as well. BST, or bovine somatotropin, is a growth hormone injected into cows. (The "r" stands for "recombinant," which means the hormone gene was cloned.) Contrary to what its name might suggest, injecting cows with this hormone simply makes them produce more milk, instead of growing into super-sized giants.

Alex was curious why rBST-free milk was such a big deal to this café owner, so he asked. A little intemperately, she responded, "Do you know where milk comes from?" Stunned by her condescension to a paying customer, Alex (a PhD in microbiology) patiently listened as she went on a diatribe about how hormones injected into the cow ultimately end up in the milk. She insisted that these nasty hormones were bad for human health—and she wouldn't be caught dead serving lattes with rBST-laced milk.

A lot of people buy this argument. Companies have heeded the protests of consumers like the crazy café lady. Containers of Ben and Jerry's ice cream carry promises that they've made every effort to avoid rBST, as do many brands of organic milk. The only problem is that everything the café proprietor said was factually incorrect.

The most obvious problem with her argument is that everything, from vegetables to meat, has hormones in it—the label "hormone free" is completely bogus. Also, if you're reading these words, you're not a cow. Humans do not respond to cow growth hormones. According to researchers at Cornell University, you could inject BST directly into your veins and nothing would happen.[36] Hormones also come in a variety of types—BST is a protein hormone. Like any other protein that you eat, BST is degraded during the digestive process. Finally, the Cornell researchers point out that BST levels are not elevated in the milk produced by hormone-injected cows. The FDA agrees with these conclusions—which should put an end to this phony health crisis over rBST.[37]

But the anti-hormone crowd was having none of it. In response to these findings, the anti-BST folks struck back, claiming that a different hormone, "insulin-like growth factor 1" (IGF-1) is elevated in BST-supplemented cows. Additionally, they claimed this hormone might cause cancer in humans. Again, the FDA responded: "Careful analysis of the published literature fails to provide compelling evidence that milk from rbGH [rBST]-treated cows contains increased levels of IGF-1 compared to milk from untreated cows."[38] The FDA went on to conclude that there was no increased risk of cancer from drinking the milk.

Undeterred, the anti-BST activists changed tactic again, now focusing their attention on the welfare of the cow. They now assert that injecting cows with BST causes mastitis (udder infection) and that this results in needless pain and suffering for the cow. It is true that supplemented cows experience a small but significant increase in their chance of acquiring an infection.[39] However, any cow that produces more milk, including ones that are not injected with BST, has a greater chance of developing mastitis.[40] In other words, normal cows are prone to the same infection if they are particularly efficient milk producers.

Did the crazy coffee shop lady know all of this? Alex, an actual scientist with real scientific data, went back and asked if she would like to read the data he found. She told him, brusquely, "No."

In a nutshell, that's the problem with the organic food movement. Those who endorse it will not change their mind in the face of overwhelming scientific evidence. On the contrary, it is common to hear activists attack the FDA for being a pawn of the biotech industry. As we have seen time and again, in issues ranging from climate change to vaccines, when activists don't like the science, they attack the scientists.

Raw Food: What You Don't Know Can Kill You

Associated with the organic food movement is a particularly troubling trend of consuming raw or unpasteurized food. One of humanity's top priorities since the dawn of civilization has been to *not* poison ourselves. The consumption of raw food essentially ignores thousands of years of common sense—and yet that's exactly what those who believe the propaganda of advocates of natural, untreated, unprocessed, organic food end up doing. After all, if it's natural, it must be good, right?

In 1996, a company called Odwalla proved that natural isn't always good. That was the year the company was held criminally liable for poisoning several dozen people and killing a young child with its unpasteurized apple juice.[41] Unsurprisingly, the un-pasteurized juice contained a pathogen, *E. coli* O157:H7. This is not the average, everyday *E. coli* that you find in poop. Instead, this is the bacterium associated with *E. coli* outbreaks in ground beef that carries with it a nasty toxin designed to kill your intestinal cells. Normally, the outcome of infection is bloody diarrhea. If you're unlucky, a more serious complication is hemolytic-uremic syndrome, which causes the death of red blood cells and kidney failure. This life-threatening complication is more likely to occur in children and is often responsible for any deaths due to food poisoning caused by *E. coli* O157:H7. Additionally, raw or unpasteurized honey may contain spores of the bacterium *Clostridium botulinum* and if fed to babies can result in infant botulism.

Surely, you would think that the threat of poisoning a bunch of little kids would put the brakes on the raw food movement, but you would be sadly mistaken. Rawism, as it is called, is still alive and well—and poisoning adults as well as kids. A particularly fashionable raw food trend is to drink unpasteurized milk. Milk, of course, comes from cows, the same cows that often carry *E. coli* O157:H7. Anyone who has ever witnessed an old-fashioned cow-milking knows it is not a pretty process. The *Seattle Times* vividly describes a raw milk producer whose cow defecated and splattered him in the face while the farmer kept milking away.[42] Is it really a surprise that *E. coli* from his farm was later found in sick patients who had consumed his unpasteurized milk?

The rejection of pasteurization is every bit as troubling and puzzling as the rejection of vaccinations. In 1938—what your grandparents refer to as the "good ol' days"—approximately 25 percent of all foodborne and waterborne illnesses were due to milk. Thanks to mandatory pasteurization, today that number is less than 1 percent.[43] The scientific and medical communities strongly support pasteurization. An FDA official warned, "Raw milk is inherently dangerous and should not be consumed by anyone, at any time, for any reason."[44] This is particularly true for children, who are not in a position to make an informed choice for themselves. Because of that, one former USDA official suggested that providing raw milk to children should be a criminal offense.[45]

The basic microbiology behind food safety is overwhelming: raw or unpasteurized food products can make you very sick. (Raw fruits and vegetables, even organic ones, are usually fine if they are thoroughly washed first.) People who consume undercooked meat and unpasteurized dairy products or juices are taking unnecessary risks with their own health. Worse, giving these products to children, whose bodies are often not strong enough to combat foodborne illnesses, is downright irresponsible and should be illegal. It turns out there is a good reason that the FDA puts warning labels on certain kinds of raw food.

Streaking Around the Neighborhood

In the realm of troubling trends, another weird manifestation of the "natural is better" myth is running barefoot. A popular book, *Born to Run* by Christopher McDougall, helped popularize the idea that running sans shoes is what humans were designed to do. Proponents of caveman-style running claim that they experience less pain and injury when they run barefoot. This bizarre phenomenon has become so popular that there is even an annual New York City Barefoot Run.

Fine. But consider the following: a professor at the University of Washington once told a story about his experience as a medical doctor working in the Amazon. The native people in the region regularly experienced infectious diseases that most Americans (including some microbiologists) have never heard of. One of them was a parasite called *Strongyloides*, found in soil and water that has fecal contamination. This particular parasite, a small worm that grows only a couple of millimeters in length, often enters the human body through the feet. It turns out that this infection is relatively easy to prevent: wear shoes. (When the natives, who lived in extreme poverty, were given shoes, they were highly prized items—so prized that the natives wore them only to church on Sundays.) Thankfully, the worm makes its home mostly in tropical areas around the world, though it can be found in the southern United States.

This anecdote should cure anyone of the silliness that running around barefoot is a good idea. Besides parasites, fungi and bacteria also live in the soil, and some of them are quite nasty. Spores from the bacteria that cause tetanus and anthrax (*Clostridium tetani* and *Bacillus anthracis*, respectively) are ubiquitous in the soil. Both of them would gladly set up shop in your feet if given the opportunity.

It is indeed strange that science has to remind people of what is essentially common sense: wear shoes when you go running. And don't stop off for any unpasteurized milk, either.

The progressive desire to return us to a Stone Age culture just because it is more "natural" is bad enough, but things get even worse when we discover progressives' anti-science resistance to the *future* of food in the next chapter.

Chapter Four

FOOD FIGHT!

*Don't Walk to the Store, Drive—
If You Love the Environment*

PROGRESSIVE ACTIVISTS love to claim that government policies are all wrong and that we aren't sufficiently heeding science when it comes to the environment—that is, until science starts to deviate from core progressive beliefs about what the environment needs. That's when science transforms from something we should trust to propaganda we should dismiss, and progressives become indistinguishable from the conservative ideologues they claim to oppose.

When it comes to environmental issues, what progressives advocate depends on which environment is under discussion. Contemporary progressivism is a "big tent" ideology. There are different specialties for each kind of activist, and each defines "environment" with a different slant. There is the animal environment, for which groups like PETA (People for the Ethical Treatment of Animals) raise money and support their own pet policies. Then there is the earth environment, where groups like Greenpeace specialize in causing trouble. Then there is the environment that grows food, the provenance of groups like the dubiously named Union of Concerned Scientists. Just as each branch of the environmental movement picks a particular focus, the science it uses to support its ideas and actions is similarly narrow. We call this "à la carte science"—picking up any scientific datum that supports a particular worldview and discarding the rest.

Let's take an issue that all three of these groups, the animal rights activists, conservationists, and food fetishists, should be able to get on board with: innovations related to the future of food production. Everybody agrees that the current system of food production could benefit from more science optimization. Industrial farms do a fine job of feeding us and keeping our grocery shelves stocked—but there are still downsides to the ways that food is currently produced. As we add more people worldwide, there will be a greater demand for all types of food, particularly meat. Traditional methods of food production will be insufficient to meet this growing need, so agriculture will have to be scientifically enhanced. And because agriculture stresses the environment, finding innovative ways to grow more food locally is both ecologically and consumer friendly.

Let's address one example of a food source that is cheap for industrialized nations but not developing ones: meat. Imagine if it were possible to take meat out of centralized farms and produce it in local laboratories. This idea may sound like science fiction, but the science to do so is perhaps just ten or twenty years away. Logically, this ought to be a home run for progressives. Lab-grown meat prevents the slaughter of animals—PETA should be delighted by that prospect. Meanwhile, this method cuts the greenhouse gases out of the process—the conservationists and climate change hawks should, reasonably, be quite pleased about that. And it does all of this while cutting the waste, additives, and externalities of industrial food down massively— thus, the Union of Concerned Scientists should be signing up to support this research to get lab meat into our diets pronto. Nevertheless, as we'll see, this is an issue on which progressives are opposing science once again. Global food production is an issue of immense importance, say progressives. We agree. Science can provide the solution to this problem if progressives would just get out of the way.

Junk Math: It Takes 1 Gallon of Gas to Produce 1 Pound of Beef

Why is it so hard for progressives to think objectively about food production? The very first issue is an innate resistance to meat—and a whole slate of à la carte science to justify this revulsion. Let's be clear: there is no question that food production creates emissions. To end up with a truckload of meat, you have to start by growing enough plants to feed a herd of animals. Because of fertilizers, growing plants adds emissions; then we have the pollutants produced by burping cows; and finally processing and transporting the meat add their own environmental stresses. The more people eat meat, the more plants are needed to feed the animals, the more trucks are needed to drive the meat to the stores, and so on.

But how many emissions *exactly* are involved in meat production? Here is where the à la carte science comes in. There are so many variables that it's impossible to calculate a believable figure, just as it is unclear how many emissions are involved in printing and mailing an environmental fund-raising brochure. Undaunted, anti-meat activists have invented a metric anyway, and it has as much bearing on reality as the claims of numerologists who attribute false significance to ordinary numbers. The statistic, which you may have read countless times in books and articles, is that 1 gallon of gasoline equals 1 pound of beef. This junk statistic has been repeated so often that it is now treated as fact. When you look scientifically at this figure, it is little different from an

episode from May 2011 when Family Radio Worldwide broad-
caster Harold Camping claimed he was able to "scientifically"
assert that the world would end starting on May 21 and that the
ending would last for five months. He asked his audience to make
a few assumptions: namely, that the numbers five, seven, and ten
had special religious meanings beyond the numerical ones and
that the universe regarded the length of an earth day as more important than anything else. (His math would not work for the
length of one day on Mars, for example.) Once you were inside
his self-contained logical bubble, he was able to prove the world
would end. Except the world didn't end.

So, first things first, let's debunk the "it takes 1 gallon of gas
to make 1 pound of beef" myth as little different from Harold
Camping's numerological prognostication. A 1995 book called
Beyond Beef: The Rise and Fall of the Cattle Culture by Jeremy
Rifkin first made this claim. How did he derive it?

We know how many calories are in 1 pound of beef: its com-
position is consistent. However, we don't really know how many
calories are in a gallon of gasoline in France versus China because
gasoline is not consistent; it varies from sample to sample. Some
gasoline contains 110,000 BTUs (1 BTU—British thermal unit—
is the energy needed to raise the temperature of 1 pound of water
by 1 degree Fahrenheit) of energy, while others contain 125,000.
Gasoline also has different BTUs in different seasons, and its final
energy value is contingent on the efficiency of the machine that
it is in.[1] Those factors alone make it hard to accept a one-size-
fits-all meat-based Doomsday scenario, but this hasn't stopped
vegetarian doomsayers, who have claimed that if you eat meat
and walk to a store, you are contributing more to global warming
than *people who drive*. A UN study called "Livestock's Long
Shadow" used that fake math and UN prestige to bizarrely claim
it was better for the environment to drive a car to a grocery store
than to walk if you ate any meat.[2] By the gallon-pound metric,

the calories your exercise required for a person to walk to the store led to more emissions than driving the car did. Chris Goodall, author of *How to Live a Low-Carbon Life*, added a dash of science to some "truthiness" to create this gem for his readers:

> Driving a typical UK car for three miles adds about 0.9 KG of CO_2 to the atmosphere. If you walked instead, it would use about 180 calories. You'd need about 100 grams of beef to replace those calories, resulting in 3.6 KG of emissions, or 4 times as much as driving.[3]

How did the inventors of this bogus equation come up with such a catchy and counterintuitive claim? They made a fundamental math error that would have got them a failing grade on a tenth-grade chemistry class: they mixed terms and guessed.

There are about 975 kilocalories of energy in 1 pound of USDA 85 percent lean ground beef.[4,5] In the case of gasoline, we have to use averages because samples vary too much, but, on average, there are 114,000 BTUs in a gallon of regular gasoline at a moderate temperature. To convert BTUs to kilocalories, divide that by 3.97 BTUs, which gives 28,715 kilocalories for an average gallon of gasoline. So, 1 pound of beef has 975 kilocalories, while 1 gallon of gas has 28,715 kilocalories—nearly thirty times more energy.

So where did Rifkin's ridiculous number come from? From a compilation he edited in 1990[6] that took it from a *Los Angeles Times* article in 1986 written by a member of the environmental advocacy group Worldwatch Institute. The only basis for this entire movement was the following sentence: "The average one-pound feed-lot steak costs: five pounds of grain based on government figures; 2,500 gallons of water (mostly through feed-crop irrigation); the energy equivalent of a gallon of gasoline, and about 35 pounds of eroded topsoil."[7] And the only actual verifiable fact in that statement was the grain. Everything else was vague conjecture, but suddenly Rifkin became the primary source for the claim. In our chapter on science journalism you will read about how the field lost its way, partly due to not using critical thinking or skepticism when it comes to pet causes; this gas-beef metric is a good example. As recently as 2009, a *Scientific American* writer again resurrected this bogus number for the audience without checking its accuracy.[8]

What's worse, this fallacious equation was taken as an axiom from which further, even crazier claims could be extrapolated. Ecologist David Pimentel of Cornell claimed a much worse scenario: that it takes 54 kilocalories of fossil fuel to produce 1 kilocalorie of beef—a whopping 1.83 gallons of gas for 1 pound of steak.[9] How was *that* figure derived? Pimentel took the 28 kilocalories as fact and then supposedly created an additional chain of fossil fuel cost for meat consumption and counted the expenditures along the way. There are so many additional assumptions that it could mean anything to anyone, except people who expect science or facts.

Ultimately, this sacrosanct statistic—that 1 gallon of gas is needed to make 1 pound of beef—is ludicrous, and that's without even considering the elaborate chain of materials that goes into making 1 gallon of gas or a car, all of which would make the math look even sillier. Walking, as common sense tells anyone who is not a progressive food activist, is far friendlier for the environment than driving.

Yet this metric keeps getting regurgitated by people who want to create a scientific reality to match their cultural worldview. What they ignore is that there is a simple physics issue to emissions—more of them are bad for us, and we want to cut emissions where we can. Forget the à la carte science and junk statistics. As the population continues to rise, we need to be innovative in coming up with ways to keep food production from becoming a strain on the environment. No question, meat is likely the biggest culprit in our emissions-heavy foodways, but meat has always been resource intensive. (Hunting was hard work too.) It is difficult to force people to change their lifestyles, but easier to encourage scientists to come up with ways to reduce or even eliminate the impact of meat production. The trouble is, progressives seem bent on fighting science every step of the way.

American Capitalism
Leads the World in Dematerialization

Doomsday and dystopian fantasies are much more eye-catching than good news stories. That's why environmentalists have spent the last several decades telling us that we are doomed. But the beautiful truth is that science and technology have been making terrific strides in reducing the environmental footprint of food production. To acknowledge the benefits of science in the recent past would be to concede that science can provide environmentally friendly solutions for future problems without a need for highly paid lobbyists or subsidies for pet ideas that have little hope of working.

If you were working on a farm in 1980, you were already raising food more efficiently compared to farmers of even two decades earlier, when doomsday progressives like Paul Ehrlich and Obama science czar John Holdren bemoaned the imminent threat of a population bomb, food riots, and the need for the creation of a "planetary regime"—a governing body with the "responsibility for determining the optimum population for the world and for each region and for arbitrating various countries' shares within their regional limits."[10] These thinkers were stuck in the same intellectual rut then as their descendants are today. This hysteria was so commonly repackaged that—without even an apology to Jonathan Swift—Charlton Heston starred in a successful movie in 1973 about the overpopulated world of 2020 called *Soylent Green*. Sorry for ruining the plot, but the predictable Big Business solution to the food crisis was eating people. It was a standard end-of-the-world fantasy that could have been written by Thomas Malthus or any of his intellectual descendants, all of whom continue to believe that science cannot solve the world's problems.

Yet in 1980, while the rest of the U.S. economy was withering amid the dual assault of stagflation and an oil embargo, farming was doing well. And it got much better. But throughout the 1980s, American farmers were forced to throw away food because longstanding subsidies and New Deal–era government price controls prevented them from reducing their prices to sell more of the food they were growing due to increasing efficiency.[11] Americans would not consume enough at the federally mandated prices, and American farmers were reliant on government subsidies rather than sales.[12]

Thus, while progressives at the time insisted we had to drastically limit food consumption (or engage in population control) to face the coming population crisis, American farmers defied this idea of "deconsumption" and instead engaged in "dematerialization." That is, they began to grow even more food on less land to try to grow food profitably.

A young person growing up on an average farm twenty-five years after the population bomb decade saw increased crop yields of 57 percent.[13] Meanwhile, relative emissions were 30 percent lower. In about that same period, the wealth of the people in India, a heavily agricultural nation, went up 140 percent.[14] The entire world was, on average, richer, less hungry, and happier. Subsidies, mandates, penalties, and other progressive economic schemes did not accomplish this dematerialization. Farmers had no reason to grow more of the food that was subsidized; they were already throwing it out because they grew too much.

No, progressive policies did not achieve more efficient food production. Advances in science and the invisible hand of the free market did.[15] Food became more plentiful with less environmental impact, and poor people got better lives. That's a win for everyone because, as Ernst Engel, the German economist behind Engel's law noted, there is a predictable result as people spend a lower percentage of wealth on basic necessities: as income rises and people spend less on food, they become more educated and more cultured. They enjoy a higher quality of life. And it wasn't just the developing world seeing this benefit: a French consumer in 2005 was also enjoying 50 percent more affluence but only using 20 percent more energy to get it.[16]

To put this consumption elasticity in context, Ron Bailey writing at Reason notes,

> If the average productivity of the world's farmers were raised to the current level of productivity of a corn farmer in Iowa, a world of 10 billion people could be fed an American diet on about half the farmland being used now. This means that an area the size of Amazonia could revert to nature.[17]

That's what science and technology can accomplish: a progressive goal of more open land, lower environmental impact, more food, and a better-educated society. Science should be a progressive's best friend.

But it is not. As soon as science disagrees with the progressive worldview, it gets chucked. For efficiency to increase, there must be a way for people to make money being more efficient. Yet because progressives are suspicious of Big Business (see the Occupy Wall Street movement of 2011), they reject science-based solutions to global food production, which often come from Big Ag and biotechnology firms. As we'll see, they poison the public's mind by introducing doubt without evidence.

A Science Solution to Growing More Food with Less Emissions Is Close

The folks at PETA crave a food system in which fewer animals are killed. Almost everyone else agrees that we would be better off with fewer carbon emissions. A food production solution that involves less greenhouse gas should be welcome to all. The good news is that a scientific solution is almost here.

It couldn't come quickly enough. The reality of food production is unromantic and unappetizing. (Have you ever visited a livestock processing plant? All mass food production is a little jarring, even a peanut butter factory—but slaughterhouses that work on an industrial scale take the cake and can spoil your appetite for days.) Yet if we want to have a world where a growing population can enjoy the same lifestyle as we do—a notion that should be the hallmark of progressive thought—we need to be more efficient. If meat is available only to the rich, the world is always going to be unequal. Moreover, it's definitely not a healthy world if meat is not available at all.

Besides, studies in Australia show *vegetarian practices* will actually result in a lot more environmental damage and even more animal cruelty than farming meat does.[18] Increased wheat and rice farming means clearing native vegetation and causing the deaths of thousands of Australian animals per hectare.

Most cattle slaughtered in Australia graze on the wild rangelands that are 70 percent of the continent. Why change land that is unused pasture into giant farms for vegetarian crops? If there were more vegetarians, that would mean either the arable land in use would need to be even more intensely farmed—more fertilizers, more herbicides, and more pesticides that will damage biodiversity and environmental health—or Australia would have to convert natural pastures where cows roam into corporate farmlands.[19]

Thankfully, science may be a few decades away from making livestock farms obsolete. One day, scientists may be mass-producing meat in a laboratory on a large scale. In vitro meat, as it is called, is a very promising technological advance. Although synthetic meat sounds nasty, it certainly isn't any worse than the process that put a hot dog on your plate.

Because the technology is still in its infancy, in vitro meat isn't hitting shelves just yet, but it already has proof of concept.[20] In 2000, researchers used goldfish cells to create fish fillets and *Time* magazine placed in vitro meat in its list of the top fifty break-through ideas of 2009.[21] Even as far back as 1932, Winston Churchill predicted, "We shall escape the absurdity of growing a whole chicken in order to eat the breast or wing, by growing these parts separately under a suitable medium."[22] We could cer-tainly use some innovations in this field—the World Health Or-ganization has projected that by 2020 meat production would be nearly double the rate it took all of human history to achieve by the year 2000.[23]

Laboratory meat would seem to be a happy confluence of sci-ence and culture for almost everyone, from vegetarians to PETA to advocates for the poor—everyone except Kansas City cattle distributors. Using a Life Cycle Assessment method in *Environ-mental Science and Technology*, researchers found that in vitro meat will cost half as much energy, reduce emissions more than 80 percent, and obviously involve almost 100 percent less land than is used now.[24]

How can anyone who claims to care about the environment and to accept science balk at meat that can be grown with fewer emissions and less land? The meat is created using lipids, amino acids, sugars, and every mineral that goes into a hamburger. It simply has to have some dye injected into it to look red and pretty—just like a hamburger grown in Texas.

But anti-science progressives deny that economies of scale will improve the process of growing laboratory meat. They claim that because amino acids and sugars have to be extracted from plants, the process can never be done more efficiently or cheaply.[25] They also don't seem to understand that a "sugar" in biology is not cane sugar, so it won't cause the economic and environmental costs of table sugar to spike—one of their more ridiculous objections. Indeed, their opposition is based on philosophy, not science: they hate meat, and they want you to hate it, too.

Instead of trying to understand basic biology and nutrition, food activists prefer to stay beneath the progressive umbrella by framing their arguments carefully to appeal to vegetarians and animal rights proponents. They once again peddle doubt without evidence: namely, that shady Big Business will secretly introduce animal-derived additives into laboratory meat. In other words, that scientists will cheat and lie—especially if they work in the private sector.

To help make sure this technology is delayed for as long as possible, food activists finance claims that the process of producing amino acids for in vitro meat will be far more expensive and resource intensive than the current system. They have even attempted to discourage scientists from working on in vitro protein by threatening years of lawsuits and FDA red tape. (Yes, perverse as it may seem, progressives use the threat of government bureaucracy as a way to scare other progressives.)

The progressive obsession with denying people food—be it genetically modified or laboratory grown—makes sense when you consider that progressives appear to love planet earth far more than other human beings.

Chapter Five

CRAPPY CONSERVATION
AND CLEAN ENERGY CHAOS

How Progressives Killed Your Toilet

WHEN IT COMES TO WATER CONSERVATION, environmentalists like to say that "every little bit counts." Turn off the water when brushing your teeth. Take showers using heads that slowly squirt out water. Reuse your towels at the hotel so that housekeeping can do less laundry. And, most importantly, make sure you replace your old-fashioned, working toilets with new models that use less water. Environmentally optimizing your next bowel movement doesn't sound like much fun, but it's the least you can do for the health of Gaia, right? So use that plunger with a smile on your face.

More and more, Americans find the latest green bathroom innovations to be irritating in the extreme. And politicians are just as ticked off. In 2011, libertarian-leaning Republican Senator Rand Paul of Kentucky castigated the deputy assistant secretary of the Department of Energy in an Energy and Natural Resources Committee hearing for this very problem. His words live on as quite possibly the best (and only) toilet-themed political rant in history:

> You're really anti-choice on every other consumer item that you've listed here, including light bulbs, refrigerators, toilets—

you name it, you can't go around your house without being told what to buy. . . . Frankly, my toilets don't work in my house. And I blame you and people like you who want to tell me what I can install in my house, what I can do. You restrict my choices.[1]

Amen, brother.

As righteous as Senator Paul's rage may have been, he picked the wrong culprit. It wasn't the Department of Energy but rather environmentalists who were moved to give us modern toilets. In 1992, they persuaded the legislature to pass George H. W. Bush's Energy Policy Act, which required that toilets use no more than 1.6 gallons per flush. Ever since then, Americans have had to live in fear, peering nervously into gurgling toilets, praying that the water goes down instead of up.

Environmentalists typically respond that such small inconveniences should be tolerated in the name of saving the planet. Remember their mantra, "Every little bit counts." But does it? Not if you do the math. Public supply and domestic water use constitutes a mere 12 percent of all water usage in the United States.[2] In fact, broken down by category, the table on page 71 shows how Americans used water in 2005.[3]

The generation of electricity by steam turbines is responsible for half of the nation's water use. Irrigation is nearly a third. Household activities (which we calculated by adding public supply to domestic use)—such as doing the laundry, washing dishes, and using the bathroom—consume only an eighth of the nation's water. Therefore, encouraging better power plant and agricultural efficiency is the most sensible approach to conserving the maximal amount of water. In fact, that very strategy is already proven to work. Thanks to better farming and power plant efficiency, the United States currently uses about 10 percent less water than it did in 1980, despite adding 70 million

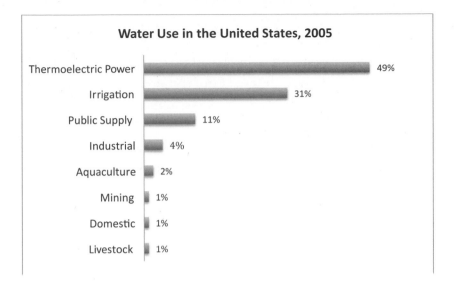

people and doubling the size of the economy.[4] Making Americans miserable with worthless toilets is worse than silly—it's bad policy.

The low-flush toilet regime has pernicious effects on the other side of the pipes as well. Low-flow toilets have been named the culprits responsible for sludge accumulating in San Francisco sewers.[5] The problem has gotten so bad that the smell of rotten eggs often wafts through parts of the city. To fight this, San Francisco is spending $14 million to dump bleach into the sewers. Bleach, which obliterates everything in its path including bacteria and viruses, is not exactly the most environmentally friendly chemical. Once again, in another misguided "feel-good" attempt to save the earth, environmentalists have managed to conjure up a civic disaster where none previously existed. Come to think of it, the image conjured up by a clogged toilet seems a rather apt metaphor for modern-day, trendy environmentalism.

Solar Power Is Good, but Only If It's Unionized

Since the nineteenth-century origins of progressivism, those under its banner have claimed to care about the environment. President Theodore Roosevelt, a progressive Republican, created the national park system. An avid hunter, Roosevelt wanted to ensure that nature was managed responsibly. That is why modern sports enthusiasts (who are often conservative) support sensible environmental regulation. They love the environment just as much as progressives do.

However, the key distinction between modern environmental progressives and the original breed is that pragmatism has been effectively eliminated from the former's political positions. That is why they oppose oil drilling in the Arctic National Wildlife Refuge (ANWR), despite the fact that 78 percent of Alaskans support it.[6] The people of Alaska know they are blessed with gorgeous surroundings; they don't really need lectures from environmentalists who live 5,000 miles away—and who have never been to Alaska, much less the part of the state closer to the North Pole than to the continental United States.

This inflexibility and unwillingness to strike a healthy balance between protecting the environment and promoting economic development are perhaps best exemplified by the ongoing fiasco surrounding solar power. Solar power rightfully excites a lot of people. The sun provides our planet with an awesome amount of clean energy, enough in a single hour to power the world for an entire year. In 2011, a company called Tessera Solar announced plans to construct an enormous solar thermal plant outside Barstow, California. Given the subject—green energy—you would think that progressives, who claim to be great champions of the environment, would greet such a development with enthusiasm. This 850-megawatt project was set to provide electricity on the scale of a coal or nuclear power plant, provide a boost to the local economy, and help reduce greenhouse gas emissions. It was as clear a win-win scenario as can be imagined. However, progressives used every legal maneuver possible to halt construction. Far from being welcomed with open arms, Tessera Solar found its path blocked.[7]

Speaking on behalf of some allegedly distraught tortoises and lizards, opponents protested that construction would damage the Mojave Desert's fragile ecosystem. Opposition wasn't uniform, however. Some progressives embraced the project, arguing that more solar power would be a net positive for the planet. However, many others were overtly hostile and challenged, "How can you say you're going to blade off hundreds of thousands of acres of earth to preserve the Earth?"[8] According to this logic, the only way to truly preserve the earth is for humans to stop building anything whatsoever.

There was a twist to this all-too-familiar story of progress versus the environment: labor unions joined in solidarity with environmental activists, ostensibly on the same grounds. CURE (California Unions for Reliable Energy) accused Tessera of not rigorously analyzing the consequences for local wildlife.[9] However, there was a self-serving catch. CURE has a history of opposing projects that are constructed by companies that do not agree to hire union workers. The *New York Times* provided some background, reporting that CURE had showed its true colors in a similar instance in 2009. Two competing companies, Ausra and BrightSource, planned to build solar power plants in California.[10] Of the two projects, BrightSource's was to be the bigger and more ambitious. However, only Ausra was on the receiving end of CURE's environmental harassment—an endless stream of legal hurdles and environmental objections. Meanwhile, CURE issued a full-throated endorsement of BrightSource. Why the difference? BrightSource agreed to hire union contractors, and Ausra did not.

This strategy of using environmental laws to extort employers is known as "greenmail." Labor unions have become adept at this technique, and at a time when jobs are scarce, places like Barstow are not amused.[11] CURE's thuggish behavior was so transparently dishonest that it even came under criticism by other labor unions. California state energy commissioner Jeffrey Byron added, "It does strain credibility when you have an organization called CURE that is concerned with the desert tortoise and wildlife habitat and turns around and disappears when a project labor agreement is signed."[12]

The greenmail proved to be too much, and eventually things fell apart for Tessera Solar. The company lost a crucial electricity sales contract with a utility company and, scrambling to stay afloat, sold off the solar thermal project to another power company. The new owner, K Road, immediately agreed that most jobs would go to unionized labor.[13] Predictably, CURE's environmental objections evaporated, and it happily announced a partnership with K Road.[14]

Jimmy Hoffa is probably smiling, wherever he is.

Greens Were for Green Energy Before They Were Against It

There was a time when green activists were in favor of green energy. Anything renewable had to be better than the carbon-spewing, mountaintop-removing alternative. That no longer seems to be the case, as energy sources that were once considered green, clean, and renewable have been steadily dismissed by progressives as bad for the environment.

Wind power is probably the most egregious example of how activists have done a complete U-turn. Once impressed by the cleanliness and promise of this power source, progressives have split ranks on the subject. Hard-core environmentalists express concern that wind turbines kill too many birds. (As authors, we have to step in and register our protest: What is with the public's fascination with birds? They are little more than disease-infested, flying tree rats—scientifically, we advise you to avoid pigeon droppings like the plague. And speaking of plague, prairie dogs— which should be called prairie rats—are notorious for harboring fleas that can transmit the bacterium that causes the actual plague. Some creatures people think are cute are actually nasty little creeps.) But since many people seem to appreciate feathered rodents, let's examine the anti-wind-power argument more closely. If, after all, wind turbines posed a legitimate threat to species' survival, it would be reasonable grounds for sending their engineers back to the drawing boards.

It's true that wind turbines do kill a lot of birds, perhaps as many as 100,000 to 300,000 every year in the United States.[15] But buildings kill a fair amount of birds, too. In New York City alone, approximately 90,000 birds are killed annually because they don't have enough sense to avoid windows.[16] Nationwide, collisions with windows kill anywhere from 100 million to 1 billion birds annually.[17] Birds are similarly deficient when it comes to avoiding automobiles and power lines. These combined kill another 200–250 million annually.[18]

But nothing compares to domestic and wild cats. Seeking sport and snacks, America's cats wipe out as many as 500 million birds every year in the United States.[19] If you have ever wondered what your cat is doing while wandering away from the house, now you know: it stalks the neighborhood, murdering birds. Still, cats don't even top the list of hazards to birds. The #1 killer of birds is habitat loss, which may destroy another 1 billion every year.[20] Simply put, it is very hazardous to be a bird in this world.

Altogether, the United States is awash in the blood and feathers of about 5 billion innocent birds every year. Therefore, every day 13.7 million birds meet their untimely demise. That sounds like an enormous number, but approximately 10 billion birds breed every year in the United States.[21] Obviously, wind turbines, buildings, ferocious cats, and habitat loss aside, they get about their business successfully and frequently.

Assuming the worst (300,000 wind-power-related bird homicides in America), we can calculate that out of 5 billion bird fatalities, a mere 0.006 percent in this country are due to wind turbines. From a bird's perspective, wind turbines are near the bottom of their list of concerns. If we truly want to help birds, there are better ways than banning wind turbines. Perhaps we should ban housecats. (Apparently Nico Dauphine, a former Smithsonian bird researcher, came to the same conclusion. In 2011, she was found guilty of animal cruelty for attempting to poison cats.)[22]

But that's just one argument against wind turbines. The other is even less scientific: locals resist new energy projects because they think wind turbines are ugly.[23] Nantucket Sound is a beautiful place that also happens to be very windy. Among its other merits, it is an ideal location for a wind farm. A project named Cape Wind was established to take advantage of this virtue and provide readily available, green energy to Massachusetts residents. It was the first offshore wind farm approved by national, state, and local governments. But there's a catch. Well-heeled locals are virulently opposed to the Cape Wind project. Forget about birds—they are concerned the distant turbines will disturb their pristine view from the beach and make yachting excursions marginally less enjoyable. For this breed of progressive, saying, "Yes," to clean energy has been replaced by NIMBYism—not in my backyard. The simple fact is that wealthy faux green activists claim to be completely in favor of renewable energy—unless they have to bear the burden.

Wind power is not the only clean energy source on which some environmentalists have turned their backs. Hydroelectric power was championed by green activists for decades. The trouble is, damming a river creates reservoirs, which entails a certain amount of plant life dying. When this matter decomposes, the process produces carbon dioxide and methane, both greenhouse gases. In what has become sadly typical of the alarmist, over-the-top nature of environmentalist rhetoric, *New Scientist* magazine reported in 2005, "Contrary to popular belief, hydroelectric power can seriously damage the climate."[24] Now it is second nature for progressive activists to dismiss hydroelectric power as damaging, even though science has since proven such claims patently false. In 2011, scientists concluded that hydroelectric power emits only a sixth of the greenhouse gases previously attributed to it.[25] A researcher from the Cary Institute of Ecosystem Studies concluded, "Hydroelectric reservoirs are not major contributors to the greenhouse gas problem."[26]

Undeterred, green activists have found other reasons to oppose hydroelectric power. Just as turbines kill birds, dams interfere with fish. Taking that logic to its unfortunate extreme, environmentalists in Ohio blocked the construction of a hydroelectric plant because it would endanger plants and inconvenience fish.[27] This tactic has gone international—greens in Brazil opposed the construction of an 11-gigawatt hydroelectric dam.[28] Who would have guessed that river ecosystems were so endangered that environmentalists would prefer sticking with dirty coal and oil?

Even solar power, perhaps the poster child of truly green technology, has a darker side. The manufacture of solar panels relies upon conventional energy sources, such as coal, and often involves so-called rare earth (a misnomer because in actuality they are not all that rare) metals. These metals are crucial to the high-tech industry and appear in everything from cell phones and laptops to solar panels and wind turbines. Extraction of these metals is costly and incredibly damaging to the environment, which is why the United States ignored its own reserves for years and let China dig them up, destroying its own environment instead.[29] Americans then happily import these metals from the Chinese, while simultaneously criticizing them for having a poor environmental record.

Rare earth extraction aside, compared to fossil fuels, solar power is still the environmentally superior choice.[30] Environmental activists have so far mostly refrained from protesting solar power (unless the panels are manufactured by non-unionized labor). But since solar is not completely clean, it's only a matter of time before environmental activists turn their backs on this technology, too.

Let's summarize the progressive environmentalists' positions on generating electricity. Wind power—bad because it is responsible for 0.006 percent of bird fatalities. Hydroelectric power— harmful to ecosystems. Natural gas—too dirty. Coal and nuclear—don't even ask. Solar power—only if you can get unionized labor to build it. Yikes, science appears to be running out of options. Tidal and wave power might work, but those, too, will disturb ecosystems. Geothermal energy can cause earthquakes. Anti-matter and fusion power sound promising, but currently those remain in the realm of science fiction.

There is a greater point to be gleaned from all this: risk-free, cost-free, pollution-free energy is simply a myth. No energy source is perfect, and every energy source either poses a risk or produces some sort of environmental disturbance. We must determine which risks are acceptable, which costs are affordable, and what level of pollution is tolerable. We must not, as the old adage goes, let the perfect become the enemy of the good. Progressive ideology when it comes to energy boils down to a kind of utopianism: progressives seem to be holding out for a miracle solution—an unlimited, completely renewable, environmentally sound energy source with no downsides whatsoever. Unfortunately, that "perfect" energy source does not, and probably will not ever, exist.

If progressive environmentalists ran the world, they wouldn't be able to face any of these tough decisions. They would nix any new project with any environmental risk at all. Then all of us would be left sitting in the dark.

Eco-Fads: Feel-Good Policies Replace Reality

Energy is not the only issue in which progressive environmentalists have replaced reality with visions of utopia. Environmentalism as a whole seems to have taken a turn toward fashionable feel-good policies that allow activists to brag about how green they are, but these policies in actuality harm the environment.

In his book *Eco-Fads*, Todd Myers meticulously details what he refers to as "trendy environmentalism."[31] Myers ruthlessly slays every sacred cow of the environmental movement, starting with LEED (Leadership in Energy and Environmental Design) certification, widely regarded as the gold standard for "green" buildings. Myers relates how environmentalists in Washington State successfully lobbied for the construction of "green" schools on the grounds that they would consume less energy and make students smarter. What actually happened? Many "green" schools consumed more energy and had poorer student achievement scores than their traditional counterparts—and they cost more to build.

Such examples of good intentions gone wrong are typical of the modern environmental movement. Local and city governments that elect progressive leadership can serve as case studies in such misguided boondoggles. In December 2008, Seattle was hit by a storm that dumped several inches of snow, an unusual occurrence in the famously mild and rainy city. Civic leaders in most of the United States know the solution to this hazard: putting salt on the street can make roads navigable by lowering the freezing point of water, thereby melting the ice. However, the progressive, avowedly green mayor of Seattle, Greg Nickels, decided against using salt, upholding a decade-long prohibition on its use due to environmental concerns. His chief of staff for the Department of Transportation asserted that excessive salt would be bad for Puget Sound—a body of water that is already salty.[32]

Instead, Seattle chose to innovate. City workers used rubber-tipped snowplows to pack the snow down, on purpose. (Apparently, regular snowplows would further damage Seattle's already disastrous road infrastructure.) Navigating the city became nearly impossible as the roads essentially turned into large ice-skating rinks, complete with "ice potholes." On top of these snow-packed roads, Seattle then poured 9,000 tons of sand in a desperate attempt to help create traction for cars.[33] It didn't work. Large areas of the city were inaccessible, even by police cars, creating an enormous public safety risk.

What was the science behind the mayor's decision? Even if we set aside the issue of public safety, was he at least correct that sand was better than salt for the environment? Not according to Todd Myers. Before writing *Eco-Fads*, he worked at the Washington State Department of Natural Resources. As he informs us, "The number one priority for forestry regulation is to keep silt out of streams. Enormous buffers are put up to reduce erosion, otherwise dirt gets into the gills of fish."[34] Additionally, there remains little reason to believe that salt—when applied only occasionally—has any long-term deleterious effect on the environment.[35] Nickels eventually saw the error in his ways, but it was too late to save him politically. Seattle's normally mild-mannered citizens were so enraged over his mangling of the situation that they tossed him to the curb in the following primary election. The Seattle "Snowmageddon" claimed its final victim. (Of course, in his place they elected an even more progressive environmentalist who promptly removed several lanes on major roads and replaced them with bi-cycle paths. Problem solved!)

An entire book could be written about the mismanagement of Seattle and Washington State and other progressive enclaves in the name of protecting the environment. Nevertheless, we will use just one more example to demonstrate how good green intentions often go wrong.

The Puget Sound region lies in an area known as the Cascadia subduction zone, in which one tectonic plate slides beneath another. This creates the potential for catastrophic geological phenomena, including earthquakes, volcanic eruptions, and tsunamis. According to experts, there may be as high as a one-in-three risk of a magnitude-9.0 or greater earthquake devastating the area within the next fifty years.[36] Indeed, in 2001, a magnitude-6.8 earthquake hit Seattle. It was a frightening reminder of how vulnerable the city's infrastructure was to such disasters, in particular the Alaskan Way Viaduct near downtown and the Highway 520 floating bridge over Lake Washington.

Given the possibility of major fatalities if either of these structures collapsed in a (quite likely) earthquake, one would think that Seattle and Washington State would act quickly to upgrade them. Alas, that has not occurred. Instead, endless political maneuvering and environmental impact studies have delayed both projects by years. Planning for the project to replace the Alaskan Way Viaduct, a major double-decker highway on the Seattle waterfront, began in 2001.[37] In 2004, an environmental impact study was released. Another one was released in 2006. Yet another appeared in 2008. And in 2009 and in 2010. Miraculously, in 2011, the "final" environmental impact study was announced. The replacement structure, an underground tunnel, will open for traffic in 2015, an incredible fourteen years after the earthquake that spurred Washington State to action. This was an abundance of environmental caution taken to its impractical extreme.

The story of the Highway 520 floating bridge, which could very well sink in a major earthquake, is even worse. Plans to replace this disaster-in-the-making started in 1997. After adjustment for inflation, the original bridge cost $245 million to build. Construction on the new bridge, which has barely begun and will not be completed until 2014, has already cost the state $400 million, and the total cost is projected to reach $4.65 billion.[38] The reason: once again, endless community consulting, political maneuvering, and requests for environmental impact studies—notably from Mayor Greg Nickels and the University of Washington—delayed the project and increased its price nineteen times over the original cost.[39]

There is no question that concern for the environment is a good thing. So is getting community approval. However, just as unions have developed the strategy of greenmailing, progressive environmentalists have acquired the tools to delay or block projects they do not like simply by asking for more community discussions, requesting more environmental impact studies, or engaging in other political maneuvers. While they claim to have the needs of the community at heart, in reality they are wasting taxpayers' money and time—and in the case of earthquake-prone Seattle, potentially putting people's lives at risk.

Progressive Apocalypse:
The Zombie of Thomas Malthus

Visions of the apocalypse are almost exclusively associated with (usually conservative) religious believers. However, progressives have their own cataclysmic vision of the future, one not filled by the wrath of God, but one filled by the wrath of Mother Earth. After millennia of neglect and abuse, humankind will eventually pay for its secular sins as human-caused catastrophes essentially wipe civilization off the planet. Just look at Al Gore's *An Incon-*

venient Truth, which paints a particularly vivid picture of the piti-less revenge the natural world has planned for us.

Whom can we thank for sowing such dour pessimism through-out the world? One Mr. Thomas Malthus, who, despite assuming room temperature in 1834, continues to live on in the hearts and minds of progressives all over the world—no matter how many times the rest of us have tried to kill him. Like a zombie crawling out of a grave, some ideas just never die.

Thomas Malthus sired the vision that the human population will inevitably grow beyond its means to survive. As a result, mass starvations and epidemics will wipe out a substantial pro-portion of humanity. Today, this cheery idea is eponymously titled a "Malthusian catastrophe." Malthus's intellectual descendants have expanded the vision to embrace not only food shortages and disease (as if they weren't enough), but also pollution and environmental damage. Nearly two centuries after his passing, Malthus has been proven wrong over and over again. Agricul-tural advances and technological innovation have greatly in-creased sustainability, and the human population has steadily increased to its current size of approximately 7 billion. It would seem that for now the Malthusian catastrophe is pure fantasy.

This reality, however, has not prevented continual scaremon-gering from progressive environmentalists. In his apocalyptic 1968 book, *The Population Bomb*, Paul Ehrlich predicted Malthusian mass starvations throughout the 1970s and 1980s. Fortunately for us (but unfortunately for him), things did not work out that way. Ehrlich's colleague John Holdren is another influential purveyor of doom and gloom. Together, they predicted in 1971 that pollution and other human activities would result in an ice age that—due to a slumping Antarctic ice cap—"could generate a tidal wave of proportions unprecedented in recorded history."[40] Then, to make sure they had all their bases covered, they also predicted catastrophic global warming and a whopping

150-foot rise in sea level.[41] (Just to put that into perspective, Al Gore tries to frighten people with only a 20-foot rise in sea level.) No question: it is tough to predict what the change in sea level will be by 2100. The world's leading authority on climate change, the UN's Intergovernmental Panel on Climate Change (IPCC), predicted a rise of approximately 200–500 millimeters (about 8–20 inches) by 2100 in its Fourth Assessment (AR4) issued in 2007.[42] Other climate scientists are more pessimistic, and it is common to read of projected sea level rises of approximately 1 to 6 feet. Al Gore's infamous 20-foot prediction is a worst-case scenario that involves the melting of the Greenland Ice Sheet, but this also is difficult to predict.[43] A rise of 150 feet, offered by John Holdren as a possibility, is an absurdity worthy of the plotline of a B-grade science fiction movie.

And yet it gets even weirder. Holdren's visions of the coming human-caused apocalypse led him to predict that 1 billion people could die from global warming by 2020. When criticized for that outrageous comment, his defenders focused on the word "could."[44] Apparently, it is okay for a scientist to make wild predictions in public as long as he calmly inserts the word could. For example, an alien *could* burst out of Elvis Presley's grave and lay eggs in your chest. Notice we said "could"—that makes it a responsible prediction.

John Holdren's scholarship then took a turn from strange and irresponsible to creepy. Once again, with his colleague Paul Ehrlich in the book *Ecoscience: Population, Resources, Environment*, he discussed methods to keep the world's population under control. After voluntary methods failed, they described the possibility of coercive sterilization.[45] The authors (thankfully) did acknowledge that such population control measures would likely not be acceptable to most societies.[46] But the sheer creepiness of describing inhumane population control policies in the first place seems to have escaped them. Responding to the controversy that inevitably ensued, Ehrlich and Holdren issued a press release denying they actually advocated such policies.

Even giving full benefit of the doubt and accepting that eugenics, for them, was just a distant theoretical possibility—especially since it hasn't been fashionable for progressives since the 1930s—we might assume that somebody with such outlandish, and regularly off-base, predictions of calamity would be relegated to the fringes of science and society. Alas, that is not the case. In 1991, John Holdren was elected to the prestigious National Academy of Sciences, and in 2009, President Obama appointed him director of the Office of Science and Technology Policy (a.k.a. "Science Czar"). It all goes to prove our point: apocalyptic rants are effectively the norm in this day and age—but they are only socially and scientifically acceptable when coming from progressives.

The World Is Not Overpopulated

Though the scholarship of Ehrlich and Holdren has been discredited time and again, their insidious ideas live on. A 2011 opinion article in the *Los Angeles Times*, which was co-authored by Paul Ehrlich's wife, Anne, argued that overpopulation was humanity's biggest problem.[47] In fact, the authors compared humanity to a cancer: "Perpetual growth is the creed of a cancer cell, not a sustainable human society."

Progressives generally accept that they have to listen to the ideas of doomsayers like Ehrlich and Holdren since their opponents are heedless of the threat of rampant population growth. But this is not the case. Human population growth will not be "perpetual," as the authors claim. Instead, it is widely believed that population growth has already begun to slow down, and by 2050 the world population will plateau at around 9 billion people.[48] Until recently, the United Nations agreed. In a 2003 assessment, using a medium-growth scenario, the United Nations projected a population of 9 billion until the year 2300.[49] (After releasing this data, the United Nations unexpectedly changed its forecast to a population of 10 billion and growing by 2100.)[50] Specific numbers aside, demographers are simply not predicting Malthusian uncontrolled growth of the human population—the data trends simply do not support it.

Sociologists have observed that there is an ebb and flow to population growth. As people become wealthier and more technologically advanced, fertility rates naturally decline. Instead of producing several offspring, couples tend to focus their resources on producing fewer, but more successful, children. Fertility rates also tend to fall as women become more educated.[51]

Modern Europe illustrates this phenomenon quite well. Currently, the fertility rate is below the "replacement rate" of 2.1 births per woman in all twenty-seven nations of the European Union.[52] The population is falling so quickly in Russia, which may lose 25 million people over the next forty years, that demographers are now in fact prophesying an *underpopulation* crisis.[53] The government there is desperately trying to incentivize population growth. Fertility rates are seen to fall for any number of other reasons, even without increased wealth. Iran, for instance, saw its fertility rate fall from 7 in 1984 to 1.9 in 2006.[54] The truth is that, contrary to what the Malthusians would claim, fertility rates around the entire world are falling dramatically. In 1970, the total fertility rate was 4.45; today, it is down to 2.45.[55]

Around the time the world hit the milestone of 7 billion people in October 2011, *The Economist* published an article about global demography that revealed some very surprising facts.[56] First, just to put the human population into perspective, 9 billion people (the projected population in 2050) could fit, shoulder to shoulder, on the island of Maui. The world isn't as crowded as we're led to believe. Second, the absolute population number appears less important than changes within the population (e.g., an imbalanced distribution of population growth or shifts in the median age of the population are what countries actually need to watch out for). Finally, only about 20 percent of the world has a skyrocketing fertility rate; the remainder is split between an intermediate fertility rate (2.1–3.0) and a low fertility rate (below the replacement rate of 2.1). As expected, the places with the highest fertility rate also tend to be the poorest, such as sub-Saharan Africa. If anything, economic and technological development in Africa may be a good strategy to both help lift the continent out of poverty and reduce the fertility rate.

When we pause to think about it, this is surprisingly pleasant news. Although counterintuitive, the fact is that prosperous people produce fewer children. In the long run, the "problem" of population growth is largely self-correcting. However, instead of accepting the good news, progressive environmentalists would prefer to reject the existing science and obsess over an unlikely Malthusian catastrophe. These pessimistic predictions lead to the toxic belief that humanity is little more than a parasite on Mother Earth. As a result, in the name of sustainability and compassion, some progressives favor reducing the global population by methods that are anything but sustainable or compassionate.

For instance, progressives insist on global access to "reproductive freedom"—contraception and abortion services. In practice, the latter policy has backfired spectacularly. Families around the world, particularly in Asia, tend to selectively abort baby girls. This "gendercide" results in perversely skewed male-female ratios, as bad as 130 to 100 in some Chinese provinces.[57] Obviously, serious consequences await a society in which one gender vastly outnumbers the other. Siyani, a small village populated by 8,000 people in India, includes so few women that men live together in an arrangement that resembles an extended family—a family, that is, without any women.[58] Some men look for wives for twenty years without any luck. As a result of selective abortion of females (or, in some places, neglect or outright infanticide), one economist suggested that perhaps 100 million women are missing from the planet.[59] The victims of policies of reproductive freedom thus tend disproportionately to be girls, causing all manner of disruptions to the social order.

In 2011, while Alex was giving a talk on the benefits of genetically modified food,[60] a bizarre question demonstrated the logical extreme of the apocalyptic progressive worldview. Two people in the audience asked if genetically modified crops were artificially increasing the carrying capacity of the human population, and, consequently, if we were simply "feeding a monster." Translation: Is it a good idea to feed hungry people if they are simply going to produce more children? This was a stunning and revealing insight into the mind of the modern progressive environmentalist: saving the earth was far more important than saving humanity.

This misplaced prioritization is everything that is wrong with the modern environmental movement. Earlier in this chapter we talked about the tough choice that needs to be made between preserving the environment and creating new solar plants, building new wind farms, and, yes, feeding the hungry. After years of debating science and policy with progressives, we are left with the dismal impression that environmentalists believe any technological progress at all (such as producing electricity and increasing agricultural output) is at odds with protecting the environment. It is not. (For confirmation, look ahead to our next chapter, on the energy alternatives available to us today.) Choosing between the planet and humanity is a false choice. We must choose both, for we have no other ethical option.

When green activists finally figure this out, then everybody will be an environmentalist.

Chapter Six

SUNSHINE DAYS AND FLOWER POWER

The Fetish for Solar and Priuses

IF YOU VISIT RURAL AREAS of Pennsylvania today, you will notice a lot of new infrastructure, much of it geared toward extracting natural gas. The new machinery isn't pretty, but you won't hear the locals complain. New industry has its benefits: the more money flows into the economy, the better the standard of living. Even people with no direct stake in the gas business—the folks who don't work for a gas company, don't own a store, and didn't sell mineral rights—understand that's how real economics works.

Outsiders often lament the new construction and wonder why we can't instead find energy in some desolate wilderness. Leave nature alone. Let hunters hunt and hikers hike in peace so that unsuspecting Pennsylvanians don't end up with flammable tap water.

The funny thing is, we already have found energy in faraway remote places where nobody lives. Progressives are even more opposed to that. The Arctic National Wildlife Refuge is 20 million acres of overwhelmingly inhospitable landscape. It sits at just over one thousand miles from the North Pole, above the Arctic Circle. Calling it a "refuge" is particularly ironic nomenclature, because nothing lives on most of it. Yet when Brad Johnson at

Grist wrote about ANWR, he cooed, "Only in the United States Senate could you find someone demand that a climate bill involve drilling in one of the last pristine places on earth."[1] Treehugger scribe Sarah Hodgdon called ANWR "a place that represents a connection to the natural world that has been lost across the rest of our nation."[2]

"Pristine" is an interesting word. The moon is also pristine. Actually, that's a great analogy because most of ANWR looks like the moon. Despite what you read in environmental marketing brochures about the munching caribou that will be displaced if an oil well is there, no caribou roam its frigid ice fields, and the environmental activists and progressives who marched in protest over the issue had never seen a real picture of ANWR. All they had seen were the parts a marketing campaign wanted them to see.

While ANWR's 20 million acres are quite large, the area desired for exploration is tiny, the size of a regional airport. A regional airport on 20 million acres is going to cause mass extinctions?

Setting that small section aside doesn't impact anyone or anything, and there has been oil found all around it. That also means there is likely to be oil underneath it, according to a U.S. Geological Survey in 1998—up to 30 billion barrels of oil and 34 trillion cubic feet of natural gas.[3] The section's remote locale seems ideal for drilling to find out just how much is easily recoverable and how many "super-fields" there are. The actual exploration area needed is small, and nothing lives there. Yet with the commencement of a public relations campaign against it, anti-science progressives insisted that scientists did not know what they were talking about and that some miraculous daisy chain of ecological disasters would occur if we were to even lay a finger on ANWR. They disregarded the fact that even the Inupiat Eskimos who live in ANWR's coastal plain and have been stewards of it for centuries wanted the government to find out about the oil there.[4] President Jimmy Carter and a Democratic Congress set aside the

North Slope of ANWR for exploration because even in the 1970s government scientists believed it held a large volume of oil. Locating it was just going to take time because the technology didn't yet exist to do so.[5]

Like people in Pennsylvania, Inupiats know having a pristine, protected wilderness may make environmentalists happy but tends to be less great for people who actually live there. Development brings wealth, and wealth leads to a better living. The marketing blitzes by opponents included pictures of caribou munching on grass and natives frolicking to show us all that was at risk—*you don't hate caribou or natives, do you?*—and for contrast they showed pictures of some kind of vaguely strip-mined-looking after-effects. The "after" pictures showed real blight. Opponents could just as easily have used real pictures of the "pristine" ANWR because that is what most of it is like anyway.

The real world tells a different story from progressive marketing campaigns. Why do Inupiat Eskimos support responsible development in their backyard? Because Prudhoe Bay caribou herds are *nine times larger* in the thirty-four years since oil was discovered there and Alaska has the strictest environmental standards in the world for oil companies.[6] As we have discussed before, anti-science progressives are well funded and they are loud, but they remain a vocal minority. Their lobbying hurts actual progress, even at the North Pole.

The similarly overhyped environmental campaign against the Alaska Pipeline in the 1970s also resulted in nothing like the ecological destruction progressive activists claimed would occur. Environmental lobbyists are instead ensuring we continue to make evil, oppressive dictators sitting on oil reserves wealthy.

Don't get us wrong: we're not "Drill, baby, drill" advocates. We absolutely want to see a transition to clean (including nuclear) energy. However, the à la carte science mentality in regard to energy policy is frustrating. Progressives love climate scientists, but they don't trust geologists who can find fossil fuels. In March 2012, the Senate again tried to open up ANWR for exploration. But though Democratic Senator Mark Begich of Alaska broke with others in his party and voted for the amendment, it failed.[7]

President Obama has repeatedly made energy independence a key talking point in his State of the Union addresses. In his third, on January 24, 2012, he promised to boost domestic oil production and reiterated his focus on renewable energy.

"Right now, American oil production is the highest that it's been in eight years," he said. "Not only that—last year, we relied less on foreign oil than in any of the past sixteen years."[8]

Well, that was true, but there is a dark reality to why it was so. The economy has been in a shambles. Fewer people employed meant less oil was used, so less was imported. Activists who insisted we need to go back to 1990s levels of greenhouse gas emissions got a jolt of reality about what that really meant. In 2011, U.S. emissions *were* at 1995 levels[9] and the American economy was, rather like ANWR, an economic moonscape—except for alternative energy, which benefited from continued government largesse.

Throughout his 2012 State of the Union speech, President Obama followed a predictable, if perverse, pattern: He lamented that oil and gas companies were receiving tax breaks but then tried to take credit for the fact that domestic energy production had risen continually since 2008—due to policies put in place by his predecessor. Then he called for more subsidies for clean energy companies despite any evidence they would provide jobs or promote energy independence at all.

Strangely, he also called for *public* lands to be used for energy production—the very thing environmentalists say is a bad idea with ANWR. Let's see if we can make sense of this: Government funding is responsible for 30 percent of American renewable energy revenue, but even a tax credit for development by oil companies is bad?[10] Public lands can't be used for oil exploration, no matter how inhospitable the region, yet public lands in the forty-eight contiguous states should be used by alternative energy businesses?

What gives?

The answer is simple. Green energy is a progressive pet project. Its adherents will support it on theological grounds, even if the costs outweigh the benefits. Again, don't get us wrong: Alternative energy *is* a good idea. People have been preaching "peak oil"—the moment that the oil supply begins to dry up, kicking off an economic disaster—since 1974. Those who worry about peak oil are bound to be right eventually, so we need to be researching a suitable replacement now. Progressives have said for decades that replacement *is* ready now: a renewable energy source that is cheaper and cleaner than fossil fuels that will take off if only we subsidize it for billions of dollars. They mean solar power, right?

No, they meant ethanol, the previous pet project of anti-science environmentalists and the politicians who solicit their votes.

Biofuel Buzz: Getting High on Ethanol

Starting in the 1980s when Senator Al Gore began extolling its virtues, ethanol could do no wrong. Every environmental group supported it. If you are not familiar with it, ethanol is essentially grain alcohol fermented from starch. Corn was advocated because America grows a lot of corn and it is relatively cheap. Basically, the kernels are ground into powder and mixed with water and enzymes, and the alcohol is "cooked off." The end result can then be blended with gasoline. Proponents claimed ethanol would be a net energy gain and that it would yield 67 percent more energy than it took to grow and process the corn.[11]

There was no real science to back up the lofty claims about how biofuels would save us if they were implemented immediately.

There was just the ever-present "miracle of capitalism" dancing in the moonlight. And the still-popular refrain for that siren song is that if we subsidize something now, it will become popular and then the corporate energy companies will eventually make it more efficient and cheaper. (Apparently, evil corporate energy companies cannot be trusted with fossil fuels, but they can be trusted with biofuels.)

As vice president during the bulk of the 1990s, Al Gore pushed ethanol endlessly, delighting environmentalists to no end. During the 2000 presidential election season, energy and the environment were twin pillars of his campaign. Ethanol was going to help save us from global warming, he claimed. Was there a science basis to it? No, there was future optimism, where you create some magical curve of research and efficiency going up and cost coming down if you mandate it right now and penalize competitors. If you ever read quaint stories from the 1950s that predicted we would all be using flying cars and living on the moon by now, you have a rough idea of how Gore and others like him imagine our energy policy. (And you also have a rough idea of how Newt Gingrich views our moon policy.)

By 2005, a Republican-controlled Congress and a Republican president agreed with the intense lobbying by environmentalists and corn farmers happy to sell more product and did as was asked: they mandated and subsidized ethanol for billions of dollars per year. The reign of clean energy could finally commence!

Except it didn't. What happened next was what almost every non-partisan energy scientist and every non-progressive economist knew would happen. Ethanol companies were suddenly using taxpayer money, not their own, so they could buy up all the corn regardless of cost, which they promptly did. This drove up global prices for corn, and the biggest negative impact was on poor people who needed corn for food. The process behind making ethanol turned out to be far more resource intensive than environmentalists claimed, which meant the cost was not coming down any time soon regardless of mandates. Worse, the production of corn ethanol released nitrous oxide, a gas with three hundred times the global warming potential of carbon dioxide.[12] Predictably, mandates and subsidies crippled the very capitalism that was supposed to optimize biofuel production. Eventually, environmental activists wised up and began lobbying to end the subsidies they had spent years trying to put in place. The new consensus replaced the old consensus, and ethanol was out, but scientifically it should never have been in.

Al Gore has nothing left to prove. He has an Academy Award and a Nobel Prize, so he has the luxury of honesty now. He admitted in 2010 that his reasoning was flawed and despite claiming a scientific basis for ethanol, it was instead simple politics. "One of the reasons I made that mistake is that I paid particular attention to the farmers in my home state of Tennessee, and I had a certain fondness for the farmers in the state of Iowa because I was about to run for president."[13] Hey, his Nobel Prize was for peace, not economics.

To be sure, not all biofuels are a bad idea. Cellulosic ethanol has potential. Using genetically engineered algae to produce biodiesel is another potentially good idea—if progressives overcome their phobia of genetics. One key consideration we need to bear in mind for energy policy is that instead of squandering billions of dollars on creating an artificial corn ethanol market, that money could have been used to fund basic research instead.

What's So Wrong with Keystone XL?

No environmentalists complain about the Alaska Pipeline today, despite a vituperative anti-pipeline campaign before, during, and after its construction. That construction began in 1975, thanks to the oil embargo instituted by the Organization of Petroleum

Exporting Countries (OPEC) the year before, but it was not easy going for the project before that. The *Washington Post* insisted oil industry propaganda was denying the substantial earthquake risk,[14] and despite a nine-volume comprehensive study of the environmental impact, the *New York Times* warned in 1973 that the pipeline would cause the caribou there to go the way of the buffalo because the noise would disrupt their mating.[15]

None of those concerns were valid then, and they aren't any more valid today. Not only has the ecology not been damaged, the areas where the pipeline runs have actually been improved environmentally and economically.[16] The Keystone XL project extension, an add-on to the original Keystone program proposed in 2005 and launched in June 2010, is quite similar. It is a $13 billion TransCanada pipeline to ship crude oil from Alberta, Canada, to refineries in Texas. The meat of the extension project is 1,661 miles long. Employment projections are inevitably optimistic and hard to quantify, but estimates of new jobs range from tens of thousands to 130,000, all while delivering from 830,000 up to 1.3 million barrels of oil per day, thereby reducing our need to import oil from crazy dictators by almost 15 percent.[17] And if we don't take advantage of that oil being so close, someone else (ahem, China) will. If government gets out of the way, that's a big win for the economy, energy independence, and even global ecology since America is far more environmentally responsible than OPEC. Yet the president who claimed to care about boosting domestic oil production and increasing energy independence has shown himself to be the biggest obstacle to both.

During virtually the entire time the Obama administration has been in office, this extension project has been sitting in limbo. State Department scientists found few environmental concerns, and local expert hydrologists agreed.[18] Scientists dismissed worries about leaks into the water supply near some parts of the pipeline by noting that oil could not go upward if there were a rupture, even if progressives chose to deny gravity. Independent analyses showed a wealth of benefits.[19] But the project was still trapped in a regulatory cycle in 2011. There was no valid environmental concern; the limbo was simply another triumph of progressive lobbying over inconvenient science.

Environmentalists still claimed, even if the claim was unsubstantiated, that the risk was considerable, just as they once had with the Alaska Pipeline. But in reality pipelines are far safer and cleaner modes of transportation than tankers and railroads. On the exceedingly rare occasions when a pipeline leak happens, often less than three barrels are spilled before the leak is stopped.[20] We know how to build safe, responsible pipelines in environmentally sensitive areas—we have done it for decades.

Siding with liberals and their desire for union jobs over progressive environmental hyperbole, President Obama finally seemed to relent in August, when Energy Secretary Steven Chu said that "energy security" concerns would lead to approval.[21] Yet by November, the Obama administration had waffled again and determined it would need to "reassess" the project.[22] Activists had been demonstrating in front of the White House, and the *New York Times* continued to publish story after story damning the entire fracking industry as environmentally unsound. Approaching an election year, and caught in a double bind of either disappointing environmentalists or appearing to be insensitive to the dire need for any lifeline to economic recovery, the administration hoped to put off the final decision until after the election.

Republicans sensed weakness and forced the administration to go on the record. In December, Congress placed a sixty-day cap on the "reassessment" to prevent the president from dodging a potent election issue. In protest, he canceled the extension outright, citing his concern over "an ecologically sensitive area in Nebraska," the Ogallala aquifer.[23] That was sure to make progressive environmentalists happy, but how scientific was this claim? Geologists knew this was instead another instance of a president putting aside science for a political worldview.

The Ogallala aquifer is massive. It runs from South Dakota to Texas and is important for agriculture in states such as Kansas and Nebraska. It is indeed "ecologically sensitive." But is it some pristine ecological paradise that would be ruined by what ultimately boils down to one newly implanted 3-foot-diameter pipe? Not at all, the entire area is positively overrun with pipes, and they haven't resulted in any harm.[24]

The Ogallala Aquifer is not the pristine wilderness it was suddenly being portrayed as. It's produced 24 billion barrels of oil itself and already contains over 24,000 miles of petroleum pipelines.[25] Nebraska alone has 20,090 miles of pipelines for environmentally "hazardous" liquids[26] so another 400 miles was not deemed to be a concern in Environmental Impact Statements.[27]

Thus, we agree with what Montana's Democratic Governor Brian Schweitzer said about the Keystone XL opposition: "Ninety percent of these jackasses that are complaining about the Keystone pipeline in Washington, D.C., one year ago wouldn't have even known where the Keystone was."[28]

Additionally, environmental concerns related to aquifer risk were addressed in an eight-volume environmental impact report by the U.S. State Department.[29] Those scientists voiced no concerns. So much for the ecological sensitivity argument. Pending any new developments on the issue (which, when they come, will surely be even more politicized), progressives on the environmental fringe won out over both blue-collar voters craving jobs and science itself.

Misconceptions About Energy Efficiency

The underlying motivations behind the progressive resistance to Keystone XL points to a larger truth: ideology trumps cost-benefit analyses. While environmentalists insist they want less dependence on foreign oil, they don't want us to produce more domestic oil—they want anything *but* oil, no matter the cost. Concerns about environmental safety or cost-effectiveness end up basically irrelevant so long as green energy initiatives can lead to press release claims about sustainability.

Electric cars are a prime example of putting activism ahead of science, as are utensils that don't work but are made from compostable, renewable corn. An electric car requires an array of battery cells, almost 500 pounds of them.[30] GM's EV1 NiMH (nickel metal hydride) battery pack weighed 1,150–1,400 pounds,[31] while the newer Chevy Volt has a lithium-ion (Li-ion) battery pack that weighs in at 435 pounds.[32] The batteries are very expensive ($3,000–4,000 is the listed cost, but even Prius hybrid owners have been given quotes of $7,000 and up,[33] so it's unclear what electric car battery replacement costs will be with labor included). Since the battery's ability to charge declines with use, it will need to be replaced at 100,000 miles. Tires need to be replaced too, activists argue. True, but electric car batteries are made from nickel metal hydride, and they share a quality with tires that isn't positive—they leak energy, about 20 percent of capacity within the first twenty-four hours.[34] Coupled with a tiny driving distance, there are also the polytetrafluoroethylene binder and nickel foam materials, which have substantial environmental consequences if electric cars actually get popular.[35]

Basically, electric vehicles are dragging a lot of extra weight made of toxic materials in order to use stored energy that combustion vehicles can instead generate on the go. Advocates claim that electric cars are better than combustion vehicles because they are "85 percent efficient"—they deliver energy to the wheels directly from the battery instead of through a combustion engine, which wastes energy as heat and also during idling and braking, putting them in the 20–25 percent energy efficiency range. Yet any cursory examination shows efficiency is not telling the whole story about whether or not electric cars are better. Stored electric power presents its own environmental challenges, for example, the fact that batteries are already responsible for a huge chunk of sulfuric acid in the atmosphere—and a lot more batteries mean a lot more carcinogenic landfills.[36]

Additionally, when you charge your electric car at night, where is the electricity coming from? "The wall" is not the correct answer. Most likely, the coal power plant down the street is providing the energy, which means that until we improve our basic energy infrastructure, electric cars aren't a solution.

Frank Didik, founder of the Electric Car Society, is practical in his assessment of electric cars: "When the cost of replacing the batteries and the cost to recharge the batteries are considered, the cost to run a small electric car (non-hybrid) is about three times more than a conventional economy gasoline car. . . . Electric cars are fun to have and drive, but you will not save money, nor will you really help the environment."[37]

The good news is that we probably *can* eventually make electric vehicles more efficient than combustion cars—but subsidizing current ones is not the way to do it. More basic research—particularly into making better batteries—is the key. Also, being able to run our country on cleaner energy across the board would boost the real efficiency of electric cars, as would better electrical storage. Yet that realistic solution isn't as popular with progressives as throwing money at inadequate, feel-good fallacies—and the best example of that phenomenon is solar power.

Solyndra and the Trouble with Solar Subsidies

Is solar power really better than oil? Of course it is, in the long run. Oil will eventually run out, and in the time it took you to read that bit about oil running out, the sun gave us enough energy to run the planet for a whole day.* Solar power also has proof of concept—plants have been using energy from the sun for a large chunk of the geological record, and they have obviously prospered, even when there were no environmentalists to lobby for them.

Progressives are right to talk about plants when it comes to solar energy, but that is where their fetish with efficiency makes so little sense. Efficiency expresses energy output compared to energy input. As great as they are, plants are nowhere near as *efficient* as gasoline-powered cars; photosynthesis is only 5 percent efficient.[38] Even our bodies are not as efficient at burning food as combustion vehicles are at burning gasoline.[39]

Essentially, solar power tries to mimic the behavior of plants using silicon. Non-organic crystals are grown artificially and then modified (but not genetically!) and doped up by modern science. When engineered correctly, a photon of light will punch a "hole" in the solar panel and drive a native electron from its home, where it can then be exploited for its energy. (We wonder if someday soon progressives will protest on behalf of electrons' rights?)

Tom Murphy, associate professor of physics at the University of California–San Diego, has an excellent primer on photovoltaic efficiency where you can learn the nitty-gritty details.[40] The bottom line is this: About 44 percent of the sun's energy is usable. But that's in a perfect world. In reality, we get 8–15 percent efficiency from solar, which is still quite good given that the sun is free.

Why isn't solar everywhere yet then? After all, America is home to the semiconductor industry, and "solar-grade silicon" is the best solution we have—all things considered—but just not efficient enough to meet our large energy demands. However, it

---

*The sun provides enough energy in one hour to power the earth for a year. If you do the math, this means the sun provides about enough energy in ten seconds to power the earth for a day.

is cheaper than other solar panel technologies and hence somewhat more affordable for mass use.

Given the opportunity and availability of silicon and our expertise in it, why would progressives, led by President Barack Obama, instead invest billions of dollars of taxpayer money on solar panels made from copper indium gallium diselenide (CIGS)? Why subsidize companies that are using less efficient technology when we could use good old silicon? The answer to that gets us into one of the fundamental problems with solar: it takes a lot of space, and that means a lot of cost.

Given the massive land areas needed for solar power, the Obama administration in 2009 sought to play venture capitalist to bootstrap a more efficient alternative to existing silicon panels. Despite daunting competition from Chinese cheap labor, administration officials decided to throw enough money around to make CIGS thin-film technology competitive with silicon by being more efficient than one kind (i.e., thin-film silicon) and cheaper than the other (i.e., polycrystalline silicon).

Why would they do that? Nobel Laureate and Energy Secretary Steven Chu was concerned that China had half of the global market for solar panels. Even though silicon was cheap, American labor was not, and as demand for panels went up, so would the price of silicon. A better product could make America more competitive. He had the full force of the federal government behind him and a willing boss listening. The federal loan guarantee program had been created in 2005 as part of the Energy Policy Act and was authorized at $4 billion, but that was not enough to win this "race" against China. The 2009 stimulus plan was the solution.[41] Venture capital companies, which had only a few hundred million dollars[42] invested in *all* clean energy technology in 2005—fifty years of claiming the solar breakthrough is "just around the corner" will make investors jaded—suddenly saw the government dumping a ridiculous $44 billion into *just solar* from 2009 to 2011.[43] Andrew Beebe, chief commercial officer for Suntech, a Chinese solar manufacturer, had been dismissive of efforts to make a better mousetrap in solar power: "You had folks who came in with the hubris to say, 'I know these guys have been working on this for 50 years. But I've got $50 million and I can blow the doors off this thing.'"[44] If several million would not do it, maybe several billion would. This is America, after all. How did that work out?

Not well. Chinese solar panels, still constructed with less efficient yet cheaper silicon by a non-union workforce in a place with few environmental regulations, got cheaper, while CIGS looked like less of a solution to a cost-conscious public. Then the price of silicon came crashing down, leaving CIGS-based companies like Solyndra unable to compete and teetering on the brink of bankruptcy, kept afloat only by government largesse. President Obama and other progressives were taken in by a sexy, feel-good technology that couldn't prove itself in the marketplace. Despite this, as late as the spring of 2010, Obama infamously called that company "a testament to American ingenuity and dynamism."[45] Then Solyndra declared bankruptcy, along with several other companies, taking with them billions of dollars in American taxpayer money.[46]

After Solyndra collapsed like a house of cards, the *Washington Post* conducted a large-scale investigation. It wrote,

> Meant to create jobs and cut reliance on foreign oil, Obama's green-technology program was infused with politics at every level, *The Washington Post* found in an analysis of thousands of memos, company records and internal e-mails. . . .
>
> The records, some previously unreported, show that when warned that financial disaster might lie ahead, the administration remained steadfast in its support for Solyndra.

> The documents . . . give an unprecedented glimpse into high-level maneuvering by politically connected clean-technology investors.[47]

During the 2008 campaign, Barack Obama criticized his predecessor for being too cozy with Big Oil. Yet when he took office, his administration immediately became too cozy with Big Solar.

A Little Physics Knowledge Goes a Long Way

Scientific expertise is curiously absent among green tech's most fervent supporters. An article in Earth Techling boasted that the Obama administration had approved twenty-seven clean energy projects that would generate about 6.6 gigawatts (GW) of power.[48] That sounds impressive if you don't understand physics. When you do the math, the average project will produce 0.244 GW of power (6.6 GW divided by 27), but a single large nuclear power plant can produce 1 GW of power. To satisfy the needs of a country like ours, green energy just can't get the job done currently. Additionally, a nuclear plant can produce a lot of energy in a small area. Solar power plants, on the other hand, require a large amount of land area.

Energy from the sun also isn't quite as available as activists claim. For most of the United States, the average home gets only a few hours per day of aggregated sunlight. On average, the total amount of sunlight might be six hours per day. This would mean 6 kWh (kilowatt-hours) per square meter of sunlight beating down onto solar panels. Because the average U.S. home uses only about 950 kWh per month,[49] it would need about six 1-square-meter solar panels if they were 100 percent efficient (6 panels x 6 kWh per panel x 30 days = 1,080 kWh, more than the 950 kWh Americans use on average per month). Great! So why aren't the homes of Americans covered with solar panels?

Because solar panels aren't 100 percent efficient. Instead, they are about 10 percent efficient (if we use the cheaper ones made from silicon). That would mean sixty 1-square-meter solar panels are needed, not six. (Sixty square meters of space are about 650 square feet.) That's not huge; it would fit nicely on the roof of your house (if we assume you own a house instead of renting an apartment). But the panels are not cheap to install.[50]

There's another problem: though the sun shines "on average" for six hours per day, it doesn't shine every day for six hours. And not every house is the "average" house. Have you been to Seattle? It's sunny there for only three months. It also isn't sunny at night, pretty much anywhere. That means that excess sunlight during the day must be stored for use during the night and on cloudy days. But there isn't a good technology for storing large amounts of electricity.

The alternative would be to rely on centralized areas where solar panels could be placed. That would require brand new transmission lines to link those solar farms to the grid—up to 20,000 miles more worth of lines. Also, consider that the energy lost in transmission is up to 40 percent, which means many excess panels will be needed just to make up for the losses.[51] And don't expect that imposing a whole new form of energy will be without controversy. In 2006, San Diego Gas & Electric Co. wanted to run power lines from a solar farm in the Mojave Desert to southern California but was blocked by environmental lawsuits. The company spent nearly three years and $125 million trying to get permission to cross just 25 miles of "protected" land.[52] The company still lost and has yet to reroute the lines. Just imagine the costs of enacting such a system on a national scale.

It is daunting to imagine the costs of these policies—especially when we consider that legislation is passed all of the time without any heed to the practicalities that energy companies face in the real world. In 2009, Senator Harry Reid introduced legislation to allocate federal lands for use in alternative energy,[53] and there are already plans to build solar power plants that will cover 8 or 9 percent of northern New Mexico.[54] Projects like these somehow manage to dodge a level of scrutiny that would otherwise be applied to things like fracking, natural gas, and Arctic drilling—scrutiny that turned all of those into national controversies. When President Obama said in his State of the Union address that he wanted to make public lands available for energy exploitation—the very thing we could not do with ANWR or Keystone XL, even where there would be no negative environmental impact—environmentalists should have been just as nervous as anyone.

Just how much land is needed for solar power? If the world were run completely on today's solar power technology, it would require solar cells to cover more land than all of the paved areas in the world combined—and that's without factoring in the power lines to carry the electricity.[55]

Thus, we need to be realistic when considering energy solutions. As Roger A. Pielke Jr., director of the Center for Science and Technology Policy Research at the University of Colorado, put it in the *New York Times*,

> The world would need to deploy in a round number one nuclear power plant worth of carbon-free energy every day for the next fifty years. Whether one expresses this magnitude in nuclear power plants, wind turbines, efficiency gains or some combination—it should be abundantly clear that the one factor we need most is technological innovation. Today's technologies are not going to do it.[56]

Basic research—not the government picking winners and losers in the private sector by subsidizing politically connected solar companies—is our best way to solve future energy needs in a responsible way. And solar will be a key part of that solution, *when it is ready*.

The obsession with all things natural leads progressives to embrace technologies that simply don't work. What happens when they reject "unnatural" technologies, which often do work?

Chapter Seven

BRING ON THE VACCINES AND VIAGRA

"Unnatural" Products Can Be Amazing

IN THE LAST SEVERAL CHAPTERS, we dispensed with the progressive myth that all "natural" things are better for you. In particular, we showed that organic food is nothing more than an elaborate marketing scheme used to squeeze every last nickel out of gullible shoppers. And we also demonstrated that the progressive sunshine fetish is rooted in an impractical attitude toward energy policy.

Now it's time to tackle this myth's equally false corollary: "unnatural" things are bad for you. Implicit in this belief is that everything we touch becomes somehow contaminated. That human intellect is not to be treasured, but to be condemned. That technological progress is not to be celebrated, but to be feared. That everything that has made humanity healthy, productive, and successful should be questioned. Modern medicine's ability to heal and extend our lives—perhaps the single greatest triumph of humankind—is now fingered as one of society's greatest villains.

Basically, humans are unnatural, a blight upon the land.

Emerging from the primordial intellectual ooze of progressive thought, the myth that unnatural things are bad stirs fear and opposition to everything from vaccines to cell phones.

Anti-Vaccination Campaign

There are not many things in life that shake our faith in the intelligence of humanity more than the anti-vaccination campaign. This movement is predicated upon outright lies, and it poses a deadly threat. The biggest anti-vaccine crusaders are loony Hollywood celebrities and progressive activists whose arguments are devoid of credible science. The very existence of an anti-vaccination movement stands as a stark example of the media's double standard. Journalists and pundits would rather heap abuse on the evolution deniers on the right than call out the most outspoken and dangerous anti-vaccine activists on the left. And the result is an outright public health crisis.

The big lie voiced by the anti-vaccine crowd is that vaccines somehow contribute to autism. Let us be crystal clear: The medical and scientific communities reject this as a complete myth. Not a single major study in fourteen years has validated the fraudulent claims made by Andrew Wakefield, the medical doctor largely responsible for this entire controversy. And, yes, the claims were fraudulent. The *British Medical Journal* concluded that the medical histories of the twelve patients enrolled in Wakefield's study were manipulated and that his experiment constituted an "elaborate fraud."[1] (Wakefield's license to practice medicine in the United Kingdom was also revoked for unethical professional conduct.)[2] Any remaining doubt about the safety of vaccines should have been crushed by this announcement in early 2011. However, the vaccine-autism myth has been around since 1998, and it will take a long time to dispel. Truly, the damage Wakefield and his lemmings have done to public health is incalculable.

Wakefield is just one person, but there are others society can blame for circulating this lie for the past decade. Lifelong progressive activist Robert F. Kennedy Jr. was a major player in this fiasco. He wrote a piece for Salon in which he linked autism to thimerosal (the mercury-based preservative that is no longer found in most vaccines). The article was so inaccurate it later had to be retracted.[3] Then there is Jim Carrey, who appropriately enough played the lead in *Liar Liar* and emerged as an outspoken anti-vaccine fanatic. We can also thank the most medically adept Playboy Bunny ever, Jenny McCarthy.[4] Credit also goes to Bill Maher, the late-night talking head who mocks religion as a mental disorder but is perfectly willing to embrace pseudoscience. Maher's relationship with reality is particularly unique. He is on the record claiming that the safety of vaccines is "not settled science like global warming."[5] He also believes vaccines may trigger Alzheimer's and that Western medicine poisons our bodies. Additionally, he has made statements that suggest he believes the germ theory of disease—one of the pillars of modern medicine—may not be accurate.[6]

Sadly, the pseudoscience claims have encouraged this movement to spread beyond a motley crew of uninformed progressive activists. Some libertarians now oppose mandatory vaccinations because they see them as an intrusion of "Big Government" into people's personal lives. Some religious conservatives oppose the human papilloma virus (HPV) vaccine on (bizarre) moral grounds, preferring abstinence education in place of vaccination. Then there's Republican Congresswoman Michelle Bachmann, who, during her presidential campaign claimed that vaccines caused mental retardation.[7] Despite all this, the root of the anti-vaccine movement remains in the fundamentally progressive myth that unnatural things are bad for you. In fact, as we stated earlier, the anti-vaccine movement seems to be strongest in places that love Whole Foods.[8] With the exception of Alaska, the states with the highest rates of vaccine refusal for kindergarten students are also among the most progressive: Washington, Oregon, and (the People's Republic of) Vermont.[9]

Other examples further illustrate this connection. Rejecting vaccination has become socially acceptable throughout all of California. In the progressive counties of Los Angeles and Santa Cruz, personal belief exemptions have pushed vaccination rates below the public safety threshold level of 90 percent in multiple schools.[10] (This is also true of some conservative counties.) In the radically progressive Bay Area, rich parents can send their children to the Waldorf School of the Peninsula, where the rate of vaccination is a mind-numbing 23 percent.[11] What would cause such a frighteningly low vaccination rate? The Waldorf School's "anthroposophical medical practice" might offer a clue. According to the school's website,

> Anthroposophical extended medicine does not regard illness as a chance occurrence or mechanical breakdown, but rather as something intimately connected to the biography of the human being. Handled appropriately it presents opportunities for new balance and maturity. The patient is seen and treated holistically as a being of body, soul and spirit. This approach integrates conventional practice with new and alternative remedies, dietary and nutritional therapy, rhythmical massage, hydrotherapy, art therapy and counseling.[12]

The upshot is that children are supposed to learn that when a person gets sick, the proper remedy is coloring with crayons and taking a warm bath. By rejecting modern medicine, this group of backward-thinking progressives is setting itself up to be ground zero of the next major measles or whooping cough outbreak—if not something far worse. Should one of these diseases break out, hot rocks and herbal tea cannot help, no matter how many counseling sessions or aura cleansings someone undergoes.

Obviously, these beliefs are very dangerous. Vaccines, along with basic sanitation and hygiene, are probably the most potent weapons we have against the microbial world. A massive global vaccination campaign eradicated smallpox, a horrible and frightening disease that, in its worst form, could kill approximately one in three people. In the United States in 1900, the top three causes of death were due to infectious disease: influenza/pneumonia, tuberculosis, and gastroenteritis.[13] Diphtheria was the #10 leading cause of death. These are statistics worth remarking upon: just a few generations ago, Americans regularly died from diarrhea (gastroenteritis) or obstructed breathing (diphtheria). However, thanks to advances in public health, infectious disease is not something most citizens of the developed world think about. Instead, we spend our time preoccupied with "lifestyle" diseases. Today in the United States, the top three causes of death are heart disease, cancer, and stroke.[14] We should celebrate this fact because it reveals the amazing triumph of modern science and public health over infectious disease.

Frustratingly, the benefits of vaccination, like those of so many other scientific advances, often are taken for granted by the general public. Indeed, vaccines have become a victim of their own success in the developed world. Most people today have never seen an infant struggling to breathe because of whooping cough or lose the ability to walk because of polio. Why? Because vaccines and successful public health campaigns have made many infectious diseases a mere afterthought.

This complacency, though, leads people to falsely believe that most infectious diseases have "gone away." They have not. Many pathogens can lurk around in the environment or inside asymptomatic people known as "carriers." When society lets its collective guard down, for instance by foregoing vaccination or failing to practice basic hygiene, the long-forgotten diseases can come roaring back. And that is exactly what we are seeing now. Modern-day outbreaks of measles and whooping cough are likely due, at least in part, to insufficient vaccine coverage.

Hating Big Pharma

The Occupy Wall Street movement, which began in 2011, was largely an anti-corporation movement. The protesters disliked the 1 percent—highly paid CEOs, politicians, and rich people— who, according to them, trampled on the opportunities of the 99 percent. In between engaging in acts of vandalism and taking dumps on the sidewalk, the protesters wanted to bring attention to corporate corruption. Many Americans sympathize with this in principle. The culprits are the typical progressive boogeymen: Big Oil, Big Ag, and Big Pharma, America's pharmaceutical companies.

So is Big Pharma really as evil as everyone thinks?

Well, they're certainly not angels. "Frenemies" may be the most appropriate word to describe them. Unethical behavior is not uncommon within the pharmaceutical industry. For instance, Merck knew, or at least should have known, that its drug Vioxx could cause heart problems. While the pain medication produced tremendous benefits for millions of people, it also killed several thousand people. The overall increased risk of heart attack was quite small, so many people experiencing chronic pain probably would have chosen to take the drug anyway.[15] However, the dishonesty with which Merck conducted itself left a very negative impression on the American public.

Sadly, that is just the tip of the iceberg. The list of indictments against the pharmaceutical industry is long and damning. For example, several large pharmaceutical companies export their clinical trials to developing countries, such as India, where they recruit poor and illiterate patients.[16] In these places, regulations for clinical trials are often more relaxed and easier to break. Enrollees, including children, sometimes fail to understand the experiment they are agreeing to. In rich countries, companies have been known to perform "seeding trials." As opposed to a legitimate clinical trial, in which the intent is to judge the safety and effectiveness of a drug, seeding trials are elaborate marketing events. Although the trial appears legitimate on its surface, the true intent is to determine if doctors will be more or less willing to prescribe the drug after the trial is performed.[17] Patient care is of secondary importance.

Drug companies also have a habit of manipulating the content of scientific journals. A common tactic is known as medical "ghostwriting," in which a drug manufacturer hires an agency to write a favorable paper about a product and to find a high-profile doctor or academic to endorse it.[18] Sometimes the academics are paid to "co-author" the paper. The articles are then published in influential journals in what is little more than a sophisticated advertising campaign. This practice became so common that the editor of the *Lancet*, a major medical journal, said, "Journals have devolved into information laundering operations for the pharmaceutical industry."[19]

None of this behavior is ethically defensible. Laws should be passed to prevent this dishonesty.

Yet despite all that, Big Pharma has its redemptive qualities. On the balance, it does far more good than harm. Vaccines, which have saved the lives of perhaps billions of people worldwide, are manufactured by Big Pharma. Anti-retroviral drugs, which help people with HIV live relatively normal lives, are made by Big Pharma. Medication to treat Alzheimer's and Parkinson's are synthesized by Big Pharma. Have you ever had a headache? Big Pharma makes aspirin. Does your tummy hurt? Big Pharma makes Tums and a little purple pill called Nexium. Trouble getting an erection? We hear there is a blue pill for that, too. (But call your doctor if it lasts longer than four hours.)

Pharmaceutical companies are also very philanthropic. As measured by total cash and product giving, six of the top ten companies that gave the most in 2009 were drug companies:

Pfizer, Merck, Johnson & Johnson, Abbott Laboratories, Eli Lilly, and Bristol-Myers Squibb.[20] Pfizer has given away Lipitor (a drug that lowers cholesterol) and Celebrex (an anti-inflammatory drug used to treat arthritis and other pain) to people who lost their jobs and became uninsured.[21] One biotechnology company, Genentech, gave a dying woman a $4,000-per-month chemotherapy drug for free after the Oregon Health Plan (Oregon's Medicaid program) denied her coverage but offered to pay for doctor-assisted suicide instead.[22]

In spite of their sins, pharmaceutical companies have, without a doubt, made our lives longer, healthier, and easier. Do they sometimes sell medicine we don't really need? Of course. But markets do not operate based on what people need; they operate based on what people want (and can afford). Because American consumers want drugs for every ailment known, drug companies provide the supplies to meet those demands. Put simply, Big Pharma will not waste its precious time and resources creating drugs you are not willing to buy. Sure, society would get along just fine without a pill to help amorous seniors, but the public demands one. So Big Pharma is happy to accommodate.

Another misplaced complaint is the utopian notion that drug companies should not profit from sick people. Progressives often express moral outrage that a company can profit by selling items that people need to survive. Using the same logic, farmers should not profit by selling food to hungry people and water companies should not profit from thirsty people. Let's spell out the situation: profit is vital for any company to stay in business. Without a profit, there is no company. Without a company, there is no medicine.

To be sure, Big Pharma is not hurting for money. According to the Fortune 500, the pharmaceutical industry was the third most profitable in 2009, with a profit margin of 19.3 percent.[23] The most profitable was network/communications (e.g., Cisco), and the second most profitable was Internet services (e.g., Google, Amazon, eBay, Yahoo, and Priceline). But while people routinely complain about the greedy excesses of Big Pharma, nobody seems remotely concerned about "Big Internet."

Perhaps counterintuitively, a reasonable argument could be made that pharmaceutical companies do not profit *enough*. In 2011, a serious drug shortage hit the United States. There are many reasons for this—such as FDA regulatory policies that stifle drug production—but one of them is clearly due to basic economics and property rights: When a drug's patent expires and generic alternatives become available, the original patent-holder often loses money.[24] Because a pharmaceutical company cannot simultaneously lose money and remain in business, it sometimes simply quits making the drug. As a result, shortages can occur. Additionally, it is the loss of money that prevents pharmaceutical companies from seeking cures to rare diseases or illnesses that afflict primarily the developing world. If a disease affects only 1 in 1 million people, there will not be enough patients for a company to recoup its cost of production. The same reasoning applies to patients in developing countries, who are mostly poverty-stricken. We must remember that drug companies are not charities. If they cannot recoup the cost of production, they will not manufacture the drug.

Just how high is the cost of production? Nobody really knows. Each drug is different, and costs can vary widely. (Besides, economics is one of those fields where a person can make all sorts of unjustifiable assumptions in order to reach whatever conclusion he wants.) The most commonly cited statistic for the cost of bringing a new drug to market is approximately $1 billion. Some say it could be as high as $4 billion to $14 billion.[25] People who do not like Big Pharma claim the cost is more like $55 million, but that is hard to believe.[26] A small academic laboratory with ten employees (including relatively cheap graduate student labor) could easily cost about $500,000 annually. So it is not difficult to believe that pharmaceutical companies, which employ thousands of people, have annual operating costs in the billions of dollars. Clinical trials, which are extremely expensive and can last years, increase research costs substantially.

Additionally, it should be remembered that pharmaceutical companies bear the financial burden not only of successful drugs, but also of all the failed candidates. Sometimes, drug candidates fail in Phase III clinical trials (during which the drug is compared to the current "gold standard" treatment), after millions of dollars have already been invested. Put simply, there is no guarantee that high-potential drug candidates will work in human patients or receive FDA approval. Regardless of the exact figure of the final price tag—which is probably impossible to determine with any believable precision—it should at least be acknowledged that the cost of research is considerable.

Yet that does not prevent progressives from decrying the greedy excesses of Big Pharma and turning to alternative medicine as a legitimate option.

Alternative Medicine: The World of Woo

The animosity directed at pharmaceutical companies that sell "unnatural" drugs combined with the progressive myth that only "natural" products are good for you has resulted in a surge of popularity for alternative medicine. The *Huffington Post*, perhaps the most popular progressive website in the world, has become a laughingstock among the scientific community because of its wholehearted embrace of alternative medicine.[27]

Alternative medicine goes by a lot of other warm and fuzzy names, such as complementary, integrative, or holistic medicine. It encompasses a wide variety of treatments, from the easy to ridicule (e.g., crystal healing and homeopathy) to the more mainstream (e.g., herbal remedies, vitamin supplementation, chiropractic, and acupuncture). Most, but not all, techniques have two things in common: they are poorly regulated and about as scientific as voodoo. According to the *Economist*, Edzard Ernst, a professor who studied the supposed health benefits of alternative medicine, found that 95 percent of treatments were indistinguishable from placebo.[28] The article went on:

> [Dr. Ernst] points out that conventional medicines must be shown to be both safe and efficacious before they can be licensed for sale. That is rarely true of alternative treatments, which rely on a mixture of appeals to tradition and to the "natural" wholesomeness of their products to reassure consumers. That explains why, for instance, some homeopaths can market treatments for malaria, despite a lack of evidence to suggest that such treatments work, or why some chiropractors can claim to cure infertility.[29]

Despite this, alternative medicine is a profitable industry, taking in $34 billion annually in the United States, which represents approximately 11 percent of out-of-pocket health care costs.[30] Shockingly, $3 billion was wasted on homeopathy, a therapy that a witch doctor might prescribe. The practice tosses basic medicine and common sense aside and replaces it with the insane premise that "like cures like." According to this concept, a person who exhibits symptoms of a disease should be treated with an extremely diluted chemical that, if taken undiluted, would cause the same symptoms. If that sounds crazy to you, homeopaths will argue it is because you are brainwashed by the medical establishment.

Unbelievably, the National Institutes of Health has an entire agency dedicated to studying this sort of quackery. Called the National Center for Complementary and Alternative Medicine (NCCAM), it was created—and packed full of woo-loving cronies—by progressive Senator Tom Harkin of Iowa. As an example of the sheer idiocy that is funded by this organization, $1 million was spent determining if strategically placed magnets relieved chronic pain.[31] Another $2 million bankrolled an "acupressure" study to determine if pushing on people's heads caused them to lose weight.[32] And despite the fact that we all love Master Yoda, $350,000 was wasted to study the "chi" life force.[33] Hard to justify, the study is.

The NCCAM was established in 1998. (Its predecessor, the Office of Alternative Medicine, was established in 1991.) Surely, we would think, after two decades of rigorous science-based research, the agency would have at least one major, revolutionary discovery to boast about. Alas, it does not. According to the Associated Press,

> Echinacea for colds. *Ginkgo biloba* for memory. Glucosamine and chondroitin for arthritis. Black cohosh for menopausal hot flashes. Saw palmetto for prostate problems. Shark cartilage for cancer. All proved no better than dummy pills in big studies funded by the National Center for Complementary and Alternative Medicine. The lone exception: ginger capsules may help chemotherapy nausea.[34]

The cost to taxpayers for finding out that quackery is indeed quackery? A mere $2.5 billion.

Senator Harkin is not pleased by the results of his pet project, but not because his quack agency has become the poster child for government waste. Instead, as Michael Specter reports in *Denialism*, Harkin is upset because the center he helped establish ended up disproving alternative medicine, not validating it as he had hoped.[35] This reveals such a fundamental misunderstanding of the scientific method (the goal of which is to test—not prove—your favorite hypotheses) that one medical doctor referred to Harkin as a public health menace.[36] Further, it provides a clear example of anti-science progressivism at its absolute worst. In a terribly misguided effort to place science-based and alternative medicine on equal footing, progressives lash out against scientists when they do not like what they have to say. Using his power and influence to advance voodoo-like beliefs—while bashing mainstream medicine—is embarrassingly ignorant behavior from an allegedly enlightened, progressive U.S. senator.

Does any alternative medicine work? Probably. Some Chinese herbal medicine is legitimate, but the vast majority of alternative medicine "works" because of the placebo effect.[37] The power of placebo is truly astonishing. Even irritable bowel syndrome patients who were told they were taking fake pills reported improvements in their health.[38] But this does not justify the practice of alternative medicine. People who fall for gimmicks and lies peddled by this poorly regulated, multi-billion-dollar industry are more often than not simply gambling with their health. Besides, they are rewarding hucksters, which leads to more dishonest marketing.

Sadly, terminally ill patients who have lost all hope are particularly vulnerable to the fraud of alternative medicine. William Nelson, a cross-dressing stage performer who claims to have eight doctorate degrees and operates out of Budapest (after fleeing the United States following charges of felony fraud), sells a machine that fires radio waves into the body.[39] The device, called EPFX, supposedly cures everything from cancer to AIDS. Although it is so obviously a scam, his company has sold more than 10,000 devices in the United States alone. Most of them went to the Pacific Northwest—perhaps the most progressive region in the country, which, as we noted, contains a disproportionate chunk of anti-vaccine believers.[40]

"Chemical" Is Not a Four-Letter Word

A famous news anchor once asked random people on the street to sign a petition banning dihydrogen monoxide.[41] He persuaded citizens to sign the ban because this chemical is everywhere in the environment and causes thousands of deaths worldwide every year. Even worse, it is odorless and tasteless, so most people are completely oblivious to its presence. Did people sign the petition? You bet. Did they know what they were signing? Probably not. Dihydrogen monoxide, otherwise written as H_2O, is the scientific name for water. (Never heard of it? Even though "dihydrogen monoxide" does follow the proper conventions for chemical nomenclature, the term is not actually used by scientists.)

This story quite beautifully exposes the inanity of the "chemical-free" movement. In the eyes of a scientist, everything is a chemical. And these chemicals almost always have scary-sounding names: sodium chloride (table salt), acetic acid (vinegar), 2-oxo-L-threo-hexono-1,4-lactone-2,3-enediol (vitamin C). Most people have a limited knowledge of chemistry and are naturally afraid of what they don't understand. That is perfectly fine, but it makes us susceptible to those who would like to profit from our irrational fears. Keep that in mind next time you are confronted with chemical-free food, cosmetics, cleaners, or any other product. Chemicals are everywhere, including chemical-free products. Just to prove this point, the Royal Society of Chemistry is offering £1 million to the first person who invents a truly chemical-free product.[42] As you might have guessed, no one has yet been awarded the prize.

The snake oil peddlers (chemical-free snake oil, of course) are capitalizing on our collective "chemophobia," or irrational fear of chemicals. Even the word "chemical" itself has a scary, almost evil connotation. But this is just another manifestation of the myth that unnatural things are bad for you.

We are not making the case that all chemicals are okay. Anyone who has ever walked into a research lab and whiffed an organic solvent can attest to that. But just because some chemicals are dangerous does not mean *all* chemicals are bad. Yet that is exactly how some politicians deal with chemicals. If there is only a modicum of evidence that a chemical might be unsafe, politicians and progressive activists see no problem in calling for extensive regulation, moratoriums, and congressional testimonies. Often, the argument made is that "chemical X should be banned until it is proven safe." This is known as the precautionary principle. The problem with that argument is that it sets a scientifically impossible standard because nothing can ever be proven safe. Besides, even when the evidence is overwhelming, anti-chemical activists will find another reason to justify their beliefs. (See our discussion in Chapter 3 on genetically modified organisms.) The latest controversy involves a chemical called bisphenol A. BPA, as it is more commonly known, is the most recent whipping boy in a long series of exaggerated chemical scare stories that include DDT, dioxin, and polychlorinated biphenyls (PCBs). (While not harmless, the danger associated with them was often overblown.) BPA is everywhere, particularly in plastics. When you enter a Whole Foods market in Seattle, you are greeted by an improperly spelled but strategically placed sign that reassures customers that its receipts do not contain "bisphenal-a." The reason the chemical upsets some people is because it has been labeled an "endocrine disruptor," which means it interferes with our hormones. Is it true? Sure, a little bit. But wheat, soy, and some fruits also qualify as endocrine disruptors.[43] Additionally, a major comprehensive review (by notoriously chemophobic Europeans) concluded that BPA was largely safe given the current data, even for newborns.[44] Little Johnny should be far more concerned about choking on a hot dog than being poisoned by his favorite sippy cup.

There are two big lessons to be learned from all this. First, just because a chemical has a particular effect in the laboratory does not mean it has a biologically relevant effect in our bodies. Toxicology is extremely complicated and nuanced. Simply being exposed to a chemical, even a dangerous one, is not necessarily a cause for alarm because dosage is the far more important factor. (A medical doctor once reminded us, "The difference between medicine and poison is the dose.") Studies in the laboratory, which are often done on mouse or human cells grown under artificial conditions, do not always yield the same information when they are repeated in research animals. Even then, studies that show negative health effects in animals do not always translate to the same outcome in humans. And even then, epidemiological studies do not always agree with biological studies. The bottom line is this: *Just because a published paper presents a statistically significant result does not mean it necessarily has a biologically meaningful effect.* For instance, if chemical X doubles your risk of disease Y, is that meaningful? It depends on what the original risk was. If the original risk was 1 in 1 million, doubling that to 2 in 1 million is hardly worth worrying about.

Second, the media do an enormous disservice to the public by reporting every study that suggests that any chemical-of-the-week is bad for you. Most likely, within a month or two, another study will show the exact opposite. By reporting every twist and turn, the media cause the public to become jaded and distrustful of the information they receive. Perhaps that is the biggest reason we live in a society full of chemophobes.

REJECTING VACCINES. Rejecting modern medicine. Rejecting chemistry. Embracing woo. These are the manifestations of the progressive myth that unnatural things are bad for you. In the next chapter, we'll show how some progressives reject one of the most fundamental necessities of biology: animal research.

Chapter Eight

PETA: PROFESSIONAL EXPERIMENTERS TESTING ON ANIMALS

Don't Make Us Come Down There and Objectify Women

PIKE PLACE MARKET IN SEATTLE is a unique shopping destination. The market sells the freshest seafood, produce, and flowers in the city. As a dynamic testament to the benefits of global trade and capitalism, it features eclectic items from all over the world— from Polish pottery to Asian incense. Tourists and locals alike flock to the market to partake of the tremendous diversity of goods and services it has to offer.

The market also has its own tradition. Customers who buy fish from the market are treated to the sight of their fish being thrown around the room by a group of burly young men who chant and sing in unison. In fact, tourists come to the market often just to see the fish being thrown. (Note to tourists: they throw the fish only if you actually buy it.) The fishmongers are entertaining, and everyone who comes has a good time. Everyone, that is, except for the animal rights group known as PETA.

PETA vociferously objects to the practice of throwing fish. Of course, the fish are dead—it is hard to imagine what ethical violation could be occurring, yet PETA is offended nonetheless.

According to them, throwing dead fish is the same as throwing dead cats or dogs, and they refer to the tradition as a "corpse toss."[1] In 2009, PETA protested at the annual convention of the American Veterinary Medical Association, which had invited the Seattle fishmongers to put on a show for them. Curiously, PETA accused the veterinarians of disrespecting marine life, and in what has become a tradition for PETA, protesters displayed their outrage by protesting as half-naked fish. (We will never object to naked mermaids, but this raises prickly concerns, such as how the protesters would have dressed if the animal of concern were not a fish but a porcupine or why it was okay to objectify live women but not dead fish.)

On issues like this one, PETA's antics amount to nothing more than mildly amusing publicity stunts. The group seems to be after headlines and donations more than any real change in how Pike Place does business. It is when PETA and other groups fight against animal research—which forms the foundation of almost every medical advance we have today—that they are truly dangerous. Groups like PETA appeal to feel-good fallacies and regularly attempt to emotionally manipulate the public with distortions and outright lies. In our opinion, their behavior is so egregious that it borders on being a crime against humanity.

Animal Rights Activists
Are a Threat to Science and Medicine

Animal rights activists are crazy. This is something we can say with all the authority of people who are not trained psychiatrists—meaning, not much. But even if you think of yourself as a fan of animal rights, by the time you are done reading this section, you will probably agree that the antics of PETA and their more extreme cohorts are, frankly, nuts.

Biomedical scientists regularly feel threatened by animal rights activists. For instance, University of Washington researchers often receive e-mail warnings alerting them that animal rights activists could be in the vicinity. Why? Because they have a history of violence. An article in *Nature* provides disturbing examples:

In February 2008, the husband of a breast-cancer biologist in Santa Cruz, California was physically assaulted at the front door of their home. In the same month, the biomedical research institute at Hasselt University in Diepenbeek, Belgium was set on fire. In the summer of 2009, activists desecrated graves belonging to the family of Daniel Vasella, then chief executive of the pharmaceutical company Novartis, based in Basel, Switzerland and torched his holiday home.[2]

Some animal rights extremists are so bad that they are considered domestic terrorists. Daniel Andreas San Diego earned a highly coveted position on the FBI's Most Wanted Terrorists list for bombing two companies involved in animal research. According to police, he actually timed a second bomb to kill officers and firefighters after they responded to the initial blast.[3] Eco-terrorists attempting to murder living humans as a way to protest dead animals is perverse logic, to say the least.

Violence is certainly the worst and most flagrant tactic used by animal rights extremists, but it is not the only one. Because health charities support medical research, protesters have waged high-profile boycotts against these charities whose mission is to fight devastating illnesses such as cancer, heart disease, Alzheimer's, and Parkinson's.[4] An organization called Animal Aid boycotted multiple charities in the United Kingdom, and one official warned that such behavior could set back medical progress by decades.[5] Professor Tipu Aziz summarized the issue bluntly when he said, "If you stop animal research you will stop medical progress."[6]

In the United States, PETA opposes reputable charities that fund animal research. According to its website, the charities PETA disapproves of include the American Cancer Society, the March of Dimes, and the Muscular Dystrophy Association.[7] PETA also threatens airlines that ship animals necessary for research, and many airlines are buckling to the pressure.[8] So keep this in mind next time you see a PETA activist: his or her mission is not just to protect cute puppies and kittens but also to block the basic medical research that could one day save your life. (Actually, PETA isn't very good at protecting puppies and kittens, either. It euthanizes almost every animal that comes into its "care." For instance, only 2.5 percent of PETA's dogs were adopted in 2011, a dismal record. You can imagine what happens to the rest.)[9]

Animal rights groups are also known to brazenly distort the truth about animal research. A common tactic is to present photographs, often decades old or taken out of context, as evidence of systematic animal abuse. In one glaring example, extremists obtained graphic photographs that depicted monkeys with skin ripped off their heads. The extremists claimed this was a direct result of animal research—that scientists were abusing the monkeys. In reality, the monkeys did this to each other. Whenever possible, social animals are housed together, rather than individually. On occasion, though, social animals engage in extremely aggressive behavior, both in captivity and in the wild. The photographs were taken as veterinarians were tending to the animals' wounds.[10] In other words, the situation was exactly the opposite of what the animal rights groups claimed. Another extremist used photographs depicting the red rear ends of rhesus macaques as evidence of animal abuse. Apparently, he was oblivious to the biological fact that the faces and buttocks of rhesus macaques are naturally colored red.[11]

If you thought that animal rights extremists were relegated to the fringes of society, you would be sadly mistaken. Cass Sunstein, President Obama's appointee as administrator of the White House Office of Information and Regulatory Affairs, has endorsed a plethora of unique policies in the past, including a ban on all recreational hunting, a phasing out of meat consumption, and the right of animals to sue humans in court.[12] Naturally, humans would have to represent their animal friends in court.[13] Who exactly would do this is unclear, unless Dr. Dolittle becomes available.

Even if we set aside the extremists, the criminals, and the most outlandish beliefs and strategies, the fact remains that animal rights activism has become increasingly mainstream. As ridiculous as they are, groups like PETA are unfortunately winning the hearts and minds of people all over the world. Forty years ago, Americans gave nearly unanimous support to animal research. Today, public support has dropped to only about half.[14] This is not only distressing but also deeply hypocritical. If a person has ever visited a doctor's office, walked into a pharmacy, or taken medicine, then she or he has benefited from animal research. Therefore, if animal rights extremists truly want to be consistent, then they should refuse all medical treatment on ethical grounds. If it is not moral to conduct animal experiments, then it is surely not moral to reap the benefits of the medical knowledge gained from them.

Animal Research Is Useful and Highly Regulated

To a scientist, the benefits of animal research are so obvious that explaining them becomes perversely difficult. It is similar to asking, "Why is science important?" How does a scientist begin to answer a question like that? As G. K. Chesterton once described in his book *Orthodoxy*, it is the obvious questions that are the hardest to answer. Yet we will attempt to do just that by first dispelling more myths peddled by animal rights activists.

A common strategy that activists use in their attempts to discredit scientists is to cite examples of poorly planned research in order to mischaracterize animal research as a whole. For instance, everyone has heard about the safety tests in which makeup and shampoo are rubbed into the eyes of rabbits. This is nothing more than a straw man argument. Even if some animal research is completely useless or ill conceived, that doesn't invalidate the vast majority that is meaningful. A recent British study supports our claim, finding that 90 percent of monkey research was justifiable.[15] Of course, 100 percent would be nice, but any large endeavor with 90 percent efficiency is the bureaucratic equivalent of a modern-day miracle.

Animal rights activists usually counter with the even more ludicrous claim that animal research is not tightly regulated. In 1966, Congress passed the Animal Welfare Act to regulate how research animals are treated, and it has been amended seven times since then, most recently in 2008.[16] Opponents point out that mice, rats, and birds are excluded from this federal law, ignoring that oversight from the National Institutes of Health, part of the federal government, covers those.[17] In addition, local and institutional regulations also exist. At universities across the country, for example, those conducting animal research are required to learn how to properly care for, experiment on, and euthanize all animals. Humane treatment is strongly emphasized.

Additionally, animal research ethics and the search for alternatives are active areas of debate within the scientific community. In 2011, a meeting was called to discuss the future of chimpanzee research in the United States.[18] Many scientists feel that chimpanzee research is unethical and unnecessary given less costly alternatives, but others claim that the animals are vital for continued research on diseases such as hepatitis C. The advocates against using chimpanzees in research appear to be winning. For instance, in late 2011 the NIH decided to suspend funding of new chimpanzee research.[19] Such discussions are not limited to the issue of chimps either. In 2010, several companies, including large pharmaceutical manufacturers, supported a European effort to reduce animal research in toxicology and safety studies.[20] GlaxoSmithKline no longer uses great apes in research.[21] Throughout the medical community, alternatives to animals are sought whenever possible. This is due not only to ethical considerations, but also to financial and regulatory ones. It is far easier and cheaper to do research in test tubes than in animals, so many scientists choose that route. Still, many experiments do require animals.

Which ones? Almost any experiment that involves genetics, physiology, immunology, or pharmaceuticals requires live animal models. If a scientist is to investigate the genetics of blindness, how is it possible to do so without using a blind animal? We can't grow eyes in the laboratory. Even if we could, it is impossible to know what the eyes are seeing if they are not a part of a living animal. If a researcher wants to elucidate the neuronal connections that make consciousness possible, how can it be done without using a living animal? We can't grow brains in a petri dish. If a doctor wants to determine the efficacy of a new human vaccine or experimental drug, how is it ethical to do so without first testing it on an animal?

Animal rights activists respond by saying that animals are not good models to study human disease. That is obviously not generally correct—but even in the cases where it is, what alternative do we have? Imagine this scenario: medical researchers think they have discovered a new drug that might cure depression. Nothing is known about the molecule other than test tube experiments, and computer models suggest it has a good chance of blocking the metabolic pathways that may be responsible for depression.

The drug has never been tested in a living animal, so the scientists have no idea what dose to use or what might happen if it were injected into a person. The drug could cure depression; it also could kill a person on the spot or result in the growth of a painful cancer. Would you volunteer for that clinical trial?

We didn't think so. Put simply, scientists experiment on animals because they think it is necessary, not because they enjoy doing so. Yet many fairness-obsessed progressives continue to oppose animal research because they believe animals are equal to humans in their intrinsic worth. That fuzzy, feel-good belief is worth exploring at greater length.

Do All Animals Have Rights (or Just the Cute Ones)?

Scientists cannot answer that question, though some have tried to contend we care more about the cute ones.[22] Rights are fundamentally an ethical issue; however, a basic knowledge of biology can dispel some of the sillier arguments made by authorities in favor of animal rights.

One such authority is Princeton University bioethicist Peter Singer. A philosopher by training, he wrote a book titled *Animal Liberation* that is widely considered to be the foundation of the modern animal rights movement.[23] Essentially, he argues that because animals are sentient creatures that can experience pain, they deserve rights. If we discriminate against animals because they are not human, he considers that to be "speciesism." He even goes so far as to compare speciesism to sexism and racism. Taking that argument to its logical conclusion means that thinking you are better than a baboon is essentially the same as a man thinking he is better than a woman.

Before we get into the weeds of this fairly staggering claim, it's worth remarking on the more outlandish claims Singer makes along the way. For one thing, he is on the record saying that, ethically, sex with animals is okay. Singer admits that it is not natural or normal, but that sex with animals "ceases to be an offence to our status and dignity as human beings."[24] Basically, Singer is willing to give a halfhearted justification of bestiality, if, of course, the goat you're making love to is enjoying himself. (Predictably, Singer also advocates vegetarianism. On this issue, he gets points for consistency: it is rude to eat something after making love to it.)

To return to the peculiar principle of speciesism, this is how it supposedly works: Most people believe that we have a duty to defend our family first and humanity as a whole second. Singer says that this ignores a middle genetic category that comes between family and species: our race. However, civilized society rejects the notion that humans should promote race. Thus, because racism is unjustifiable, so is speciesism. According to Singer, promoting your species above any other species is just as immoral as promoting your race above any other.

But Singer's logic is not based on solid science. Many scientists believe that there is no real genetic distinction between races. Alan Templeton, a biologist at Washington University in St. Louis, says,

> Race is a real cultural, political and economic concept in society, but it is not a biological concept, and that unfortunately is what many people wrongfully consider to be the essence of race in humans—genetic differences. Evolutionary history is the key to understanding race. . . . There's nothing even like a really distinct subdivision of humanity.[25]

According to Templeton's research, 85 percent of genetic differences are due to individual variation, while 15 percent are due to race. In regard to this, Templeton adds:

The 15 percent is well below the threshold that is used to recognize race in other species. In many other large mammalian species, we see rates of differentiation two or three times that of humans before the lineages are even recognized as races. Humans are one of the most genetically homogenous species we know of.[26]

In other words, race is more of a cultural construct than a biological reality. Singer's argument that race is a middle genetic category between family and species is incorrect, and thus his entire argument about speciesism falls apart, purely on scientific grounds.

It is true that, even though we are one species, studies of race can be medically useful. For instance, some genetic diseases are more common among particular races, and different ethnicities can respond differently to certain drug therapies. But as the era of "personal genomics" matures—wherein a person's genetic profile can be determined quickly and cheaply—the concept of race in medical settings may become irrelevant.[27] Thus, as a whole, race is not a biologically justifiable classification for humans.

What about Singer's belief that because animals are sentient and experience pain, they deserve rights? Here again, an incomplete knowledge of biology renders this belief vague and nearly meaningless. Animal sentience and nociception (ability to feel pain) is not well understood at all. Dogs and cats clearly have the ability to think and feel pain, but what about sea sponges? They are biologically classified as animals, but they do not have a nervous system (or a digestive system). As we move up the chain of being, what about parasitic worms? Do they have rights? Singer gives us an insight into how he might answer that question:

What I'm really concerned with is the capacity to suffer and of course it's difficult to say where to draw the line between beings who can suffer and beings who can't. . . . So, give them the benefit of the doubt where you can, but in the end I'm not as concerned about insects as I am about vertebrates who I'm sure can suffer.[28]

Aha! Singer probably doesn't care about parasitic worms. But isn't that a form of speciesism? He believes vertebrates are worthy of his tender love and care, but not other animals. Singer is engaging in the typical hypocrisy that hallmarks the animal rights movement: to activists, the word "animals" usually refers only to the lovable ones, like dogs and monkeys. But activists conveniently ignore that parasitic worms and mosquitoes are also animals, thereby leaving them undeserving of activist sympathy. In reality, by "animal rights," progressives really mean "cute, fuzzy mammal rights." But isn't that inherently speciesist and hypocritical? Isn't the argument put forth by Singer and others like him illogical and self-contradictory?

Finally, it is worth discussing the claim made by Singer and other vegetarians that it is unethical to eat animals. We'll set aside the 3.8 billion years of evolution that allowed humanity to climb to the top of the food chain. Vegetarians actually kill millions of animals every year, too. When a farmer rides his combine through the fields harvesting crops, how many little mice and bunny rabbits are smashed like pancakes? A lot. In fact, so many little mammals are killed in this process that Steven Davis of Oregon State University believes that eating beef would actually result in far fewer animal deaths.[29] Regardless of whether or not he is actually correct, there are other uncomfortable facts about biology to consider: animals eat each other. There are documented cases of dogs eating their dead owners.[30] (Think about that next time Fido has a hungry look in his eye.) Carnivorous plants, such as Venus flytraps and pitcher plants, eat animals. Pitcher plants attract unsuspecting animals—including insects, frogs, lizards, rats, and even birds—into vase-like structures and slowly dissolve them.[31] In Singer's world, the basic truths about biology all pose ethical dilemmas. Indeed, Singer's world is a strange one.

Yet these kooky beliefs are becoming more popular. At various universities around the country, there is a budding academic field called "animal studies."[32] (No, it's not zoology. Think "gender studies," but replace women with chimpanzees and ponies.) And just to give you an idea of how mainstream the idea of "species equality" has become, consider this: when the president of the Humane Society made a joke that dogs should be referred to as "Canine-Americans," the media actually took him seriously.[33]

Animals and People
Are Fundamentally Different

In July 2011, a demented man named Anders Breivik detonated a car bomb and shot several dozen people at a youth camp in Norway. All together, more than seventy innocent people were ruthlessly murdered in a senseless act of terrorism. Most people would agree that such events represent humanity at its worst. But not all would agree. At a concert in Warsaw, singer Morrissey commented, "We all live in a murderous world, as the events in Norway have shown. . . . Though that is nothing compared to what happens in McDonald's and Kentucky Fried Shit every day."[34] In Morrissey's world, eating hamburger and chicken is far worse than murdering human beings.

Extreme progressives like Morrissey (and to a slightly lesser degree, Peter Singer) believe in the equality of all life on earth, and as a result, they do not view humanity as anything special. We disagree. Yes, humans are primates, but something separates us from all the other animals. Without resorting to religious or spiritual arguments, we can make a strictly scientific case that humanity is superior to all other life on the planet. But like most things in biology, this is not a simple black-or-white issue. The characteristics that make humanity unique are largely based on quantity, not quality. In other words, even though we share many of our traits with other animals, the scale and magnitude of those traits are far greater in humanity than in any other life form.

Marc Hauser, who resigned from Harvard University in 2011 amid controversial allegations of research misconduct, wrote an eloquent essay explaining the fundamental cognitive differences between humans and animals.[35] According to Hauser, there are four key traits that contribute to what he calls our "humaniqueness." First, *generative computation* allows humans to create an unlimited variety of expressions, including everything from music to words. Think of the multitude of songs that have been sung over the millennia or the fluidity of the English language, which blesses us with such creative expressions as "Octomom" and "dingleberry." Second, *cognitive promiscuity* allows us to combine ideas from unrelated fields. For example, a dash of molecular biology, a pinch of paleontology, and a heaping spoonful of fiction gave us *Jurassic Park*. Third, *mental symbols* allow us to express real or imagined experiences in written, spoken, or artistic form. Fourth, *abstract thought* allows humans to ponder everything from geometry to God, while the thoughts of the most brilliant animals are largely rooted in sensory perceptions. Our unique ability to idealize existence and dream of perfect worlds forms the basis of our morality.

Of course, animals exhibit some of these traits to very minor degrees. Animals have a very primitive form of communication. A paintbrush-wielding chimpanzee displays enough creativity that it can fool some people into thinking they are looking at praiseworthy works of modern art.[36] (Yes, that's a real experiment—though it may actually demonstrate more about "modern art" than about the creativity of chimps.) Additionally, as Hauser described, some animals appear to weigh future options when making decisions.[37]

Pigeons, for example, like to gamble for big food payoffs.[38] Though (like in humans) this actually demonstrates a tendency to make bad choices, the behavior also suggests that pigeons may weigh risks and benefits when making decisions.

As impressive as animals are, however, an enormous chasm remains between the mental abilities of humans and those of all other animals. The dumbest person you know is still a genius compared to the smartest chimpanzee.

In addition to advanced cognitive functioning, interspecies altruism also could arguably be considered a uniquely human trait. Examples of intraspecies altruism do abound in nature. Bacteria commit suicide when growth conditions are unfavorable in order for their genetically identical sisters to survive.[39] Sterile worker bees and ants labor tirelessly to support the reproductive success of a single queen. Many other animals feed and protect weak members of the group. But with the major exception of humanity, interspecies altruism is a rare phenomenon. Dolphins have been observed helping stranded whales,[40] and chimpanzees will help humans retrieve sticks.[41] And, as we all know, smart dogs can be trained to fetch beer from the refrigerator. (Okay, maybe that isn't technically altruism.) But these examples are uncommon.

Animal rights groups fail to comprehend this fact. They ironically claim that animals deserve the same rights as humans because they are just like us, but they are not. When members of PETA demonstrate against fur coats, they are engaging in behavior that no other animal on the planet would do: lobbying strongly for the welfare of another species. Simply put, animals are not nearly as concerned about us (or each other) as we are about them.

In no way does this suggest that animals do not deserve our protection and affection. But it is in the best biological interest of humanity to promote the welfare of our own species first and foremost. Answering the needs of other humans comes before answering the needs of other animals. That is why animal research is justifiable. Or as comedian Dennis Miller once wrote,

> If we're talking essential medical research that will save human lives, well, I don't give a rat's ass about . . . a rat's ass. You know, if it's between my heart or a gorilla's . . . sorry, Koko. It's been nice signing with you.[42]

He's right. Most people correctly believe that humanity must properly steward the planet, but we are still superior to all the other animals. Therefore, as long as animal research is conducted compassionately and ethically, we must engage in it as a necessary means to improve the human condition.

Sadly, real animal rights abuses occur all over the world. Poaching threatens majestic rhinoceroses and elephants. Habitat loss threatens everything from lizards to gorillas. Indeed, our planet may be in the midst of a sixth mass extinction because of the careless ways in which humanity has acted.[43] Focusing on the needs of a growing human population while emphasizing sustainable usage of our natural resources—including animals— will be vital to the health of the biosphere. PETA and other animal rights groups could play an enormously helpful role in achieving this, if only they valued humans as much as they value other animals.

Instead, they use naked women to protest dead fish in Seattle.

Chapter Nine

WELCOME TO THE ANTI-RENAISSANCE

Progressive Europe Is Scientifically Backward

IF YOU REALLY WANT TO GET progressives' hearts thumping, tell them that you just got back from Europe. They usually perk up, their eyes gleaming with a distant yearning for the utopia that lies just across the Atlantic—the closest thing earth will ever have to a secular Promised Land. And, in truth, Europe is a nice place. European countries have good food. They have clean, efficient mass transit, including high-speed trains. They have an enormous social safety net and universal health care. What's not to like? Europeans are probably smarter and more pro-science than Americans, too.

At least that's how progressives think. In regard to science, it is true that Europeans are more comfortable with evolution than Americans, and it can be reasonably argued that human embryonic stem cell research is more acceptable in Europe than in the United States. (Some countries in Europe, notably the United Kingdom, Belgium, and Sweden, have very permissive laws regarding hESC research.)[1] Yet evolution and hESCs are not the only two subjects in all of science.

In reality, there is a dark side to "pro-science" Europe. Ideas about stem cells and Charles Darwin aside, Europe has put into practice many of American progressives' pet causes. It is now the policy of many European governments to viciously oppose genetically modified crops and nuclear power. These and other anti-science beliefs are so pervasive in the politics of the European Union (EU) that science writers Frank Swain and Martin Robbins of the *Guardian* newspaper blogged their concern about it:

> In 2008, vice-presidential candidate Sarah Palin drew criticism for her absurd beliefs—from Creationism to climate denialism—but few seem to realise how far that same anti-science sentiment has crept into our own politics.[2]

Adding insult to injury, in a 2012 interview with *Science Insider*, the very first European Union science adviser, microbiologist Anne Glover, said,

> If you take people's opinions, for instance by looking at the Eurobarometer, people seem to be reluctant to accept innovative technologies. They are suspicious almost just because it's new. . . . This leaves the door open for pressure groups which are against certain things and have a very loud voice. There should be more communication about the rewards of the technologies.[3]

Because Europe has such a rich history and diverse culture, many Americans believe it is also a land of tremendous education and cutting-edge technology, filled with intellectuals, scientists, artists, and philosophers. Essentially, progressives idealize Europe as a futuristic version of the United States. But that's not true. On the subject of economics, the *Economist* has summed up the current plight of Europe as a sort of "genteel decline."[4] The same term applies aptly to the science of Europe as well.

So let's take a closer look at the claim that Europe is more pro-science than America. As a giant social experiment in progressive politics for the past several decades, Europe has seen generous entitlements, relaxed social values, and a flat-out rejection of religion—in other words, all the hallmarks of a progressive utopia. This is particularly the case in the United Kingdom, France, and Germany. Because many progressives view themselves as the ultimate champions of science, it would be reasonable to predict that Europe should be the world's greatest bastion of research and technology. And yet it is not. Indeed, in many ways Europe is a continent that is looking backward, not forward.

What went wrong?

Precautionary Principle Paralysis

What is commonly referred to as the precautionary principle can be best summarized by the cliché "better safe than sorry." In Europe, this principle is law of the land.[5] Before a new technology is allowed to enter the market, society must take pains to ensure that it is completely safe. On the surface, this idea seems appealing enough. Who doesn't want their lives to be as risk free as possible? Progressive advocates, particularly environmentalists, have advocated that the same principle be advanced in the United States to put the onus on, for instance, chemical companies to prove that their products don't cause awful diseases. However, though the idea sounds good in theory, the European example makes for terrible public policy. Why?

The unspoken assumptions behind the precautionary principle are deeply flawed. First, the principle assumes that new technologies can be proven "safe." This isn't just difficult—it verges on the impossible, since science can't account for every single possible exigency. It is possible, however, to demonstrate that something is unsafe. Thus, the burden of proof should belong to the person making that claim. Second, the principle assumes that older technologies are safer than newer technologies—essentially, that we should be on alert for "dangerous" new innovations. It is this flawed reasoning that primarily motivates the organic food movement, which falsely believes that ancient farming techniques are healthier for people and the environment than modern agriculture. Never mind that using genetically modified crops is far less damaging than spraying a field with insecticide. Finally, the principle assumes that there is no risk to *not* implementing new technologies. When life-saving drugs have potentially dangerous side effects, society can be rightfully concerned about those risks, but rarely does anyone get concerned about the risk of *not* releasing the drug. What if a drug could cure cancer, but it greatly increased the likelihood of heart disease? Blocking its progress would be equivalent to condemning cancer patients to death. In this way, the precautionary principle actually retards progress, and it prevents society from advancing.

It's no mystery why the precautionary principle *seems* like a reasonable idea. People are notoriously poor at evaluating risk and probability. (How else can we explain the existence of heavy smokers who play the lottery?) As Michael Specter discusses in his book *Denialism*, the risks of everyday life are a much greater threat to our safety than most new technologies.[6] We worry that cell phones will give us cancer (despite the lack of evidence), yet we drive in our cars every day without thinking twice about the fact that thousands of Americans die in car accidents every year. The mere act of getting out of bed each day puts our lives in jeopardy. (Actually, staying in bed puts them in jeopardy, too.) The real question should not be whether risk exists (because no matter what, it surely does), but what level of risk we are willing to tolerate.

What are the practical consequences of this unhealthy obsession with risk aversion? Technophobia and the rejection of revolutionary, and possibly life-improving, technologies. And the precautionary principle is just the tip of the iceberg. The birthplace of both the Renaissance and the Enlightenment is in fact openly hostile to some of the greatest technological advances in energy, agriculture, and communication.

In March 2011, a catastrophic 9.0 earthquake hit Japan and also triggered a massive tsunami. This double disaster caused the meltdown of the Fukushima Dai-ichi nuclear power plant. A combination of old equipment, negligence, and bad luck appears to have been responsible for the meltdown. Predictably, antinuclear activists seized the opportunity to mount a campaign about the evils of nuclear power. The strange thing is that European officials actually listened. On the other side of the planet, on a continent that does not face the same constant threat of natural disasters as earthquake- and tsunami-prone Japan, Europeans were turning their backs on nuclear power.

Like a series of dominoes, country after country rejected the most powerful (and carbon-free) source of energy currently available to humanity. Switzerland, which currently gets about 40 percent of its energy from nuclear power, decided to abandon the technology over the next couple of decades.[7] Not to be outdone, the chancellor of Germany, Angela Merkel, promptly announced the closure of all nuclear plants by 2022.[8] Up until that point, Merkel had supported nuclear power. Her politically motivated decision was made even more frustrating by the fact that she holds a PhD in quantum chemistry. If a physical-sciences-trained leader is unable to stand up for science, then the future of nuclear power in Europe is indeed quite bleak. To his credit, the sex-crazed prime minister of Italy, Silvio Berlusconi—probably in between bunga bunga parties—expressed an interest in expanding nuclear power in his country. However, a nationwide referendum abruptly ended the plan when a stunning 94 percent of voters rejected it.[9]

Ironically, the decision to stop building nuclear power plants arguably makes Europe less safe. Newer plants can draw upon the lessons of Fukushima and implement improved backup systems that are better equipped to handle massive disasters.[10] Older plants, obviously, use older technology. By choosing to keep these plants in operation during the phaseout instead of replacing them with newer, safer ones, Europe might actually be taking a riskier path.

But logic is not a strength of progressive, anti-nuclear protesters. In November 2011, thousands of protesters demonstrated against a train carrying nuclear waste from France to Germany. Some of the protesters chained themselves to the rails, and others tried to sabotage the track.[11] Nuclear waste is highly radioactive, and people's concerns about it are understandable. However, actively trying to make the transport of nuclear waste dangerous simply defies human logic. What if the protesters had derailed the train, spilling some of the dangerous substance? Would Europe have been safer then?

Though their logic often fails them when it comes to public policy, anti-nuclear activists at least have enough sense to cash in on other people's fears when they see an opportunity. The former science and technology spokesman for the United Kingdom's Green Party, Dr. Christopher Busby, accused the Japanese government of deliberately spreading radiation around the country following the Fukushima tragedy. However, besides being a doctor, he seems to have been a bit of an entrepreneur. While advocating against the dangers of exposure to nuclear material, he sold mineral supplements on a California-based website to guard against this conspiracy. Japanese people could purchase these supplements at the not-so-modest price of £48 (nearly $75).[12] Scientists quoted by the *Guardian* condemned the pills as useless and accused Dr. Busby of being ignorant of basic radiobiology—not to mention alarming the public and profiting from the results.[13]

Nuclear power isn't the only victim of the precautionary principle. Europeans (and many Americans) are also opposed to "fracking," a hydraulic fracturing process where wells are injected with water, sand, and minor amounts of chemicals to create small cracks in rocks deep underground in order to allow natural gas and oil to flow up for extraction. The technique has been used since the 1940s to extract oil and natural gas,[14] and it has become so advanced that it has been used in over 1 million wells just in the United States.[15] Fracking has vastly expanded the world's energy supply, allowing access to resources that were previously thought to be unattainable. For Europe, tapping its own energy resources would mean beginning the process of freeing itself from dependence on Russian natural gas. (Russia is not exactly the friendliest or most trustworthy country in the world. In 2009, Russia turned off its gas supply to Ukraine over politics.)[16] Natural gas is also cleaner than coal, so you would think that fracking would be a win-win technology. Yet many Europeans choose instead to fixate on the evershrinking risk of environmental contamination, while ignoring the political and environmental risk of leaving an important energy supply untapped.

The fear and paralysis created by the precautionary principle also extend to agricultural policy. Europeans love organic food, despite the 2011 *E. coli* outbreak that killed several dozen people and sickened about 4,000 others.[17] European organic farmers were probably not responsible for the outbreak (because the contaminated seeds were traced to an Egyptian supplier), but the incident should demonstrate that the words "organic" and "locally grown" are not synonymous with "safer." Yet the technophobic mentality encouraged by the precautionary principle has led Europe to embrace organic food while almost completely rejecting genetically modified food. As we have discussed in earlier chapters, there is absolutely nothing to fear from GM food and much to gain. However, the continent remains stubbornly opposed to this revolutionary technology. *New York Times* columnist Thomas

Friedman poked fun at this in a column aptly titled "Ah, Those Principled Europeans":

> Europeans, out of some romantic rebellion against America and high technology, were shunning U.S.-grown food containing GMOs—even though there is no scientific evidence that these are harmful. But practically everywhere we went in Davos, Europeans were smoking cigarettes—with their meals, coffee, or conversation—even though there is indisputable scientific evidence that smoking can kill you.[18]

If that were the extent of European silliness, that might be something to laugh about. But these anti-science beliefs are quite damaging to the world's poorest people.

The European Union is notoriously slow in approving GM crops for cultivation, though not for any scientific reason, but mostly because its unscientific bureaucrats and citizens are afraid of them. As of 2011, only two GM crops had been approved for cultivation, compared with ninety in the United States and twenty-eight in Brazil.[19] One of the two crops approved, MON 810, is a variety of corn produced by Monsanto that contains Bt toxin. (A refresher, if you skimmed over our earlier explanation of Bt toxin: This insect-killing protein is harmless to humans but kills insects. It decreases the need to spray fields with environmentally harmful insecticides—and though Europeans shun the GM version, it is otherwise considered safe enough that even organic farmers spray it on their crops!)[20] Despite this, progressive environmental groups, such as Greenpeace and Friends of the Earth, helped convince the conservative government in Germany to ban the GM corn.[21] France, which had already banned the product, now has to work extra hard to keep its ban in place after the European Court of Justice ruled that the country had failed to prove that GM corn was a risk to human health or the environment.[22]

Just in case you were starting to feel that the European Court of Justice represented a voice of sanity on the continent, in 2011 the court declared that honey had to be labeled "genetically modified" if a trace of pollen from GM corn could be detected in it.[23] Essentially, if a couple of bees wander into a field of GM corn, and an almost infinitesimally small fraction of the honey is made from this pollen, the entire batch is considered "contaminated"— a useful insight into the technophobic, precautionary-minded bureaucrats who govern European institutions. The deep-rooted fear of GM technology engulfing Europe resembles an obsessive-compulsive patient struggling to touch a bathroom doorknob for fear of germs. (Actually, such a patient is probably much more rational than European lawmakers in the grip of the precautionary principle.)

Put plainly, Europe's opposition to GM technology harms poor people in developing countries the most. African governments are eager to export food to the European Union, but since they fear GM bans, many are reluctant to introduce the technology. This, in turn, helps keep African farmers poor by preventing them from competing in the global marketplace. Enraged by the injustice, Kenyan scientist Dr. Felix M'mboyi vented, "This kind of hypocrisy and arrogance comes with the luxury of a full stomach."[24]

Europe's technophobia gets even worse. In 2011, the Council of Europe—a powerful political organization that influences, but does not make, legislation—passed a resolution that could only be described as a pseudoscientific rant against electromagnetism.[25] Its bizarre recommendations included a ban on WiFi and cell phones in classrooms to protect children; consideration for the needs of "electrosensitive" people who supposedly suffer illnesses when exposed to electromagnetic fields; and credence given to "early warning" scientists.

These recommendations were made in direct contradiction to existing medical and scientific evidence. First, electromagnetic fields from wireless devices do not cause cancer. How do we know? Because basic physics tells us it is impossible. The types of electromagnetic radiation that can cause cancer, such as X-rays and UV light, have energies greater than 480 kilojoules per mole—enough to break molecular bonds and cause mutations in DNA. The radiation from cell phones is only 0.001 kilojoule per mole.[26] Second, there is no such thing as "electromagnetic hypersensitivity." A systematic literature review found no evidence that exposure to electromagnetic fields triggers the symptoms associated with the illness.[27] In fact, experiments have found that exposure to fake electromagnetic fields can trigger symptoms, suggesting a psychosomatic phenomenon known as the nocebo effect. (The nocebo effect is like the evil twin of the placebo effect in which the health outcome is harmful, not beneficial.) Finally, the recommendation to listen to early warning scientists is nothing more than a flimsy excuse to dismiss mainstream science in favor of the ramblings of crackpots. There are scientists who believe HIV doesn't cause AIDS. Is Europe interested in heeding their advice, too?

Unreasonable fear of electromagnetism explains why the European Union banned the use of X-ray scanners in airports in 2011. It is true that, in general, avoiding X-rays is a good idea. But there are two scientific reasons that the X-ray scanner ban is based on faulty logic. First, the very act of flying in an airplane might increase a person's risk of getting cancer because of greater exposure to cosmic rays. The amount of radiation a person gets from an X-ray scanner is comparably small, equivalent to a few minutes' worth of what he or she will be exposed to by flying.[28] Thus, if the EU's own logic were consistent, airplanes should themselves be banned. Second, all of this fear over radiation is based on the questionable notion that all doses of radiation are unhealthy, even extremely small ones. This is known as the "linear no-threshold model," and its validity is still debated within the scientific community. An example illustrating why this hypothesis might be wrong is the observation that cancer deaths in Denver are lower than in many other parts of the country despite the fact that people living there are exposed to more radiation.[29] (The greater dose of radiation is from living at a higher elevation and leakage of radon—a decay product of uranium—from the soil.) Additionally, the biological effects of radiation are so complicated that some evidence indicates that low-level doses might actually have some beneficial effects.[30]

Perhaps the most egregious example of how the precautionary principle warps logic and is then used as a bludgeon by progressives is the prosecution of six seismologists in Italy who incorrectly concluded that the city of L'Aquila was in no danger from earthquakes.[31] Tragically, in 2009 an earthquake struck the city and killed 309 people. Like an angry mob in the Dark Ages, the citizens turned their wrath against the scientists, and the local prosecutor charged them with 309 counts of manslaughter. Their trial began in September 2011 and will likely continue beyond 2012 with appeals, as they each face more than a decade in prison. Patrick McSharry, head of Oxford University's Catastrophe Risk Financing Center, compared the trial to a "medieval witch-hunt."[32] Thus, the precautionary principle, which encourages the unrealistic expectation that life should be completely risk free, could ruin the careers and lives of six scientists who were merely asked to provide their professional opinion on a complicated matter.

As these examples demonstrate, the precautionary principle plays on people's fears of change and uncertainty. The outcome of this in Europe has been quite clear: the principle encourages the rejection of new technologies (e.g., nuclear power) by exploiting people's poor sense of risk perception. It promotes older, inefficient technologies (e.g., organic farming) by promoting nostalgia and idealizing supposedly simpler times. It encourages paranoia (e.g., fear of wireless communication) by exploiting the public's misunderstanding of technology. And it has resulted in the prosecution of scientists who were simply performing their civic duty.

Europe, the land of progressive values, has become the land of the Luddites.

Ugly Bananas? Those Were Once Banned in Europe, Too

Surely, we must be joking. Alas, we are not. Charles Wheelan, in *Naked Economics*, writes,

> In November 2008, the European Union acted boldly to legalize . . . ugly fruits and vegetables. Prior to that time, supermarkets across Europe were forbidden from selling "overly curved, extra knobbly or oddly shaped" produce. This was a true act of political courage by European Union authorities, given that representatives from sixteen of the twenty-seven member nations tried to block the deregulation while it was being considered by the EU Agricultural Management Committee.[33]

The British referred to this scientifically dubious regulation as the "bendy banana law." But in the words of a British politician, this law looked "positively sane" compared to an EU regulation issued in November 2011 that banned drink manufacturers from claiming that water prevented dehydration.[34] Unbelievably, it took a panel of twenty-one scientists three years to reach this head-scratching conclusion. And if a drink manufacturer dared to violate the law, it could result in a two-year jail sentence. Fortunately, media and public ridicule caused the EU to reaffirm the health benefits of water.

One of the hallmarks of progressive ideology is a fixation on legislating ideology, whether through regulation, the precautionary principle, or outright bans. For every complex problem facing humankind, the progressive solution is simple: more rules and larger bureaucracies. The above examples illustrate the pitfalls of this approach. Even though some government regulation is necessary, it is not the solution to all—or even most—problems. The progressive European Union has thus morphed into an oppressive leviathan, striving to regulate even the irrelevant minutiae of its citizens' lives. In such a political atmosphere, both science and common sense are among the first casualties in what inevitably becomes a bureaucratic war on reason.

America, Not Europe (or China), Leads the World in Scientific Research

Because progressives often claim to be the ultimate champions of science, we would expect Europe—that grand social experiment in progressive politics—to be the global hub of research and technology. But it's not. The *Economist* recently lamented the fact that there is no British version of Bill Gates or Silicon Valley. It suggested the reason was because of "national and European regulations and a tepid climate for entrepreneurs."[35] Tepid, indeed. An enormous problem for Europe is the cost of labor, which consumes more than 70 percent of research and development (R&D) funds. Compare that to the cost of labor in the United States (45 percent) and Asia (30 percent, a figure that excludes Japan).[36]

But this is not the only problem Europe faces. Despite routine whining from progressives about what they perceive to be inadequate science funding from the federal government, the United States spends more money on R&D than any other nation.

U.S. Research and Development Share of Gross Domestic Product: 1953–2008

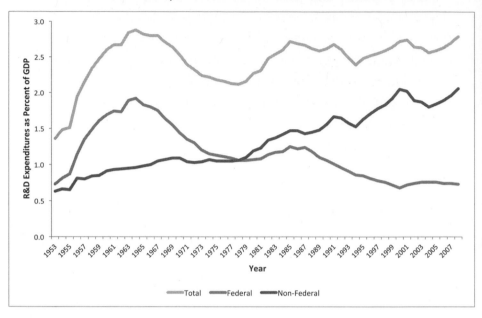

Progressives correctly indicate that, as a percentage of gross domestic product (GDP), the federal share of R&D expenditures decreased over the past several decades. But this is only half of the story because, during the same time interval, private sector contributions increased. Combined, total R&D spending has remained relatively constant over the past thirty years, never dipping below 2.3 percent of GDP, as the graph above indicates.[37]

It is important to note that federal R&D expenditures increased during the conservative Reagan administration, only to be cut massively by the moderately progressive Clinton administration. In fact, Reagan, who is routinely dismissed by progressives as a terrible, anti-science president, issued one of the greatest defenses of basic research funding for science ever:

Science has grown, and with it, the fascination it holds for all of us. But as the pursuit of science has become ever more nationally and even multinationally funded, it has also become more expensive. The problem here is that science, unlike a bridge or an interstate highway or a courthouse, has no local constituency. Today, when we're witnessing some of the most exciting discoveries in the history of science, things similar to the breakthroughs associated with Einstein, Galileo, and Newton, Federal funding for science is in jeopardy because of budget constraints.

That's why it's my duty as President to draw its importance to your attention and that of Congress. America has long been the world's scientific leader. Over the years, we've secured far more patents than any other country in the world. And since World War II, we have won more Nobel prizes for science than the Europeans and Japanese combined. We also support more of what is called basic research; that is, research meant to teach us rather than to invent or develop new products. And for the past 40 years, the Government has been our leading sponsor of basic research.

The remarkable thing is that although basic research does not begin with a particular practical goal, when you look at the results over the years, it ends up being one of the most practical things government does.[38]

When Ronald Reagan left office, federal R&D funding constituted 1.18 percent of GDP, but by the time Bill Clinton left, federal research dollars had been slashed to a mere 0.68 percent of GDP. George W. Bush, who was routinely demonized by progressives as an anti-science president, actually increased R&D funding modestly. When he left office, the federal government contributed 0.73 percent of GDP to R&D funding.[39] That included a doubling of the budget for the NIH (a process begun under Clinton)[40] and a boosting of the funding for NASA (which had seen budget cuts under Clinton).[41]

The typical rebuttal from progressives is that some other countries outspend the United States when R&D funding is compared as a percentage of GDP. That claim is used to distort the magnitude of American R&D investment. In 2012, the United States dedicated 2.85 percent of its GDP to R&D expenditures. When they are framed only in terms of percentages, a handful of countries outspent us, for example, Japan (3.48 percent), Germany (2.87 percent), South Korea (3.45 percent), Sweden (3.62 percent), Switzerland (3.0 percent), Israel (4.2 percent), Finland (3.8 percent), and Denmark (3.08 percent).[42] But consider the immense size of the U.S. economy. Even though "only" 2.85 percent of GDP went to R&D, that translated to roughly $436 billion. No other country on earth comes even close to that amount. In fact, all of Europe combined spent only $338 billion in 2012.[43]

To fully grasp just how much money the United States spends on R&D, imagine this: if all the R&D funding in the world in 2012 were put into a large pot, about a third of it would come from the United States. The pie chart on page 157 beautifully illustrates this point.[44]

There are two very important points to be made from the chart. First, the United States contributes far more money to R&D than Europe does. It is a goal of the European Union to dedicate as large a share of its GDP to R&D as the United States, but in light of the global recession and Eurozone crisis, that is unlikely to occur anytime soon.[45]

Second, the constant handwringing and lamenting over the imminent technological dominance of China is largely hype. The United States outspends China by roughly 2.5 times. Also, although much has been made over China's increasing scientific output,[46] it must be put into proper context: China has a unique tendency to cheat and steal. There isn't really a nicer way to put it. Plagiarism is an enormous problem in the Chinese scientific community. According to *Nature News*, China's "scientific journals are filled with incremental work, read by virtually no one and riddled with plagiarism."[47] It became so bad that China put

Share of Total Global R&D Spending

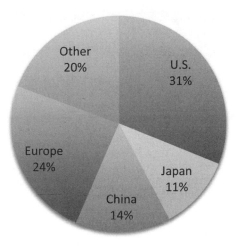

forth plans to terminate weak journals. Additionally, Chinese education prioritizes repetition (rote learning), rather than creativity.[48] This might be good for learning math or taking the standardized tests progressives use to claim American education is "dismal," but it isn't terribly useful for innovating or contributing to science in a meaningful way.

America Dominates the World
in Higher Education

Another way to measure a country's scientific impact is to determine how many of the world's top universities are located there. (This is because the frequency with which faculty members are cited by academic publications is often a criterion used in ranking universities.) *U.S. News & World Report* publishes such a list annually. In 2010, the United States had thirty-one of the top one hundred universities, far more than any other country or region.[49] (Please see graph on page 158.)

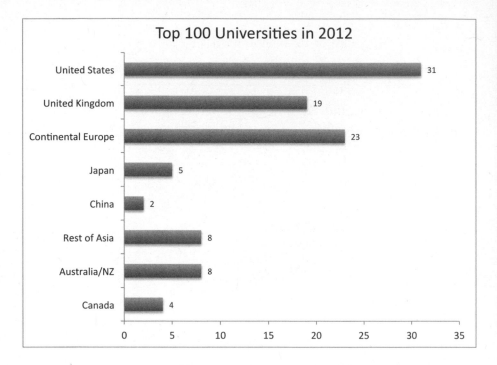

There are several noteworthy points here. First, despite persistent complaints—not just from progressives, but from conservatives, too—that American education is underfunded, we remain the best country in the world for higher education. By far. No other country is even really close.

Second, the British have an incredibly strong record in higher education. When we consider that the United Kingdom has a population a fifth the size and an economy about a seventh the size of those of the United States, possessing nineteen of the world's top one hundred universities is an outstanding accomplishment. Of course, the United Kingdom has a long, rich history of higher education. The University of Cambridge, now ranked #1 in the world, was founded around 1209. Harvard, now ranked #2, was founded in 1636—obviously, by British subjects. So the British have had several centuries longer than Americans to build their educational tradition. (Also, notice that the English-speaking world has sixty-two of the top one hundred universities. Despite the flaws of the British Empire—of which there were many—this is a phenomenal testament to its legacy. It also stresses the importance of speaking English in an increasingly globalized world. Indeed, one author wrote that native English speakers are so advantaged that it is like being born with the ability to do algebra.)[50]

Third, continental Europe has a respectable showing with twenty-three universities in the top one hundred. (Of particular note, Germany has five; the Netherlands has four; France and Denmark have two each; and Italy, Greece, and Spain have none.) But if progressives were the true champions of education, as they routinely claim, then we would expect continental Europe to dominate the world in higher education (especially when we consider how long it has been around). But it doesn't. The relatively conservative United States does.

Finally, China has a long way to go before it can challenge anyone in higher education. (In the chart, Hong Kong universities are listed under "Rest of Asia." However, if those schools were listed under China, it would raise its number of top one hundred universities from two to five.) Until its universities emerge on the global stage, it will be difficult for China to achieve scientific and technological dominance. Simultaneously, it needs to work on improving the science literacy of its population as a whole, which one survey concluded is a paltry 3 percent. By comparison, American science literacy is 28 percent.[51] That number is certainly nothing to brag about, but it is still nine times better than China's.

PUTTING THIS ALL TOGETHER, we can see clearly that the progressive paradise of modern-day Europe is not the hub of scientific and technological activity we would expect if progressivism truly benefited science. On the contrary, the center-right United States continues to dominate the world in science and higher education, and we will likely do so for the foreseeable future. That doesn't mean, however, that we should become complacent. China and other Asian countries are catching up, and Europe could, too, if it got its fiscal house in order. But for now, the United States remains #1. And in part, that is thanks to the increasing number of women who are joining the ranks of the scientific community.

Chapter Ten

BOYS HAVE WEE-WEES
AND GIRLS HAVE HOO-HOOS

Science Must Investigate Uncomfortable Issues

IMAGINE TWO CHILDREN, one boy and one girl. They live near each other and play together with an enormous Lego city. The girl likes to be in charge of the hospital, the boy in charge of the police station. A regular day consists of Lego police officers being shot by pirates—pirates live near police stations, as every kid knows—and the wounded officers are healed by Lego doctors and nurses.

How can these two children be so different? The question is far more controversial than it should be. Say, for instance, that the boy liked to pull the legs off ants and fry caterpillars with a magnifying glass, while his female friend liked to play with dolls more. Why? It's not like he had received bug-torturing lessons from his parents. The two children existed in each other's worlds on an almost-daily basis, yet their personalities and desires remained quite different. Anecdotes from childhood don't make for definitive scientific theories—yet even so, a girl who likes to heal things and a boy who likes to destroy things hints at the possibility of a greater truth: boys and girls may be inherently different.

You might be thinking, "Thank you, Captain Obvious. Of course, boys and girls are different." But is this obvious to everybody? Merely suggesting that there might be biological differences between men and women cost Harvard president Lawrence Summers, the ultimate progressive insider, his job—and fellow progressives were the ones after him.

Progressives Bring Down
the President of Harvard

At an economics conference in January 2005, Lawrence Summers made a career-ending mistake: he discussed a politically sensitive topic without adhering to progressive dogma. On considering why there were fewer women than men in science academia (particularly math- and physics-intensive professorships), Summers discussed some controversial—yet scientifically plausible—hypotheses: (1) women were more likely to choose family over rigorous academic jobs; (2) there was an inherent biological difference in scientific ability between men and women; (3) girls were conditioned and socialized in such a way that as adults they did not like science; and (4) institutional discrimination prevented women from advancing in science.[1] Whether rightly or wrongly, he felt that the evidence indicated that the first two hypotheses were the greatest contributors to the "gender gap" in science academia, whereas the latter two only played a minor role.

Unbeknown to him, Summers had just committed professional suicide, despite the fact that he was a prominent member of the supposedly diverse, tolerant world of higher education. His second hypothesis, which suggested a fundamental biological difference between men and women, particularly irked progressives. They insist that discrimination is the sole reason that fewer women are in science than men. Further, they tend to accuse anyone of sexism who merely suggests that men and women might have different natural abilities. By directly challenging that belief, Summers roused an angry mob that would not rest until Summers lost his job. And that's exactly what happened. Despite issuing unnecessary apologies that fell on deaf ears, Summers resigned as president of Harvard in February 2006. (There is also speculation that this controversy cost him the job of Treasury secretary in the Obama administration.)[2]

Summers learned the hard way that the feel-good fallacies of progressive thought are stronger than the values of free inquiry and the primacy of the scientific method. This problem has become so bad that some scientists are afraid to talk about their research for fear of being labeled sexist—just for pursuing certain hot-button topics. Though they ostensibly champion freedom of speech on college campuses, progressives actually wield the controls of a frightening censorship machine that seeks to professionally destroy anyone who dares to disagree with their cultural agenda.

We have created a culture in America in which some of our most intelligent citizens are afraid to share their knowledge with the world for fear of losing their jobs and having their reputations smeared. An article in *Slate* describes the concerns expressed by behavioral neurobiologist Larry Cahill at a panel discussion:

> His colleagues are so afraid of being called "neurosexists" that they've refused to study or acknowledge differences. This anxiety about lending credence to sexism was manifest on the panel. . . . Fear of sexism has produced a bias against conceding sex differences, which gets in the way of frank discussion and investigation.[3]

Two female scientists in the audience correctly objected that ignoring gender-based differences was anti-scientific and prevented a more complete understanding of mental illness in women.[4] However, some equality-obsessed progressives do not share this scientific attitude.

That is a shame because investigating the biological differences between men and women is a legitimate area of scientific inquiry vital to fully understanding human health. If gender differences do in fact exist, shouldn't we know about them? And how can we know about them if we are not allowed to talk about them? Clearly, the progressive value of gender equality has become perverted from its original intent. Gender equality originally meant that men and women should be treated fairly. But now this honorable idea has been twisted into the scientifically absurd belief that men and women are also biologically equal. We are not. And a mountain of scientific data exists to support this assertion.

Are Boys Smarter Than Girls?

Was Lawrence Summers right? Are boys biologically programmed to be more proficient at math and science than girls? Progressives are infuriated that such a question could even be asked in the twenty-first century. However, asking uncomfortable questions is sometimes a scientist's job. And if scientists are supposed to ask such questions, then society should be prepared to accept unpleasant answers. Regardless of how intelligence research makes people feel, the real issue is whether or not the claims are scientifically valid. So are boys smarter than girls? The available evidence indicates the answer is an unsatisfactory "maybe," but that is why the matter needs to be researched.

Summers's argument was a little complicated. He did not say outright that men were smarter than women. His statement was much more obtuse:

It does appear that on many, many different human attributes—height, weight, propensity for criminality, overall IQ, mathematical ability, scientific ability—there is relatively clear evidence that whatever the difference in means—which can be debated—there is a difference in the standard deviation, and variability of a male and a female population.[5]

Did you catch that? Read it again. Summers said that the mean (average) intelligence between men and women is debatable, but the standard deviation (variability) is clearly different, as the graph below indicates.

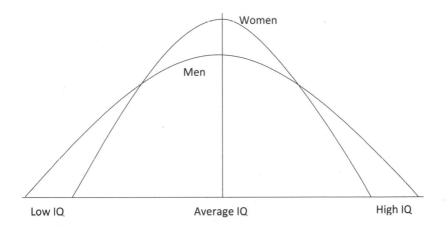

Put simply, he is implying that average male and average female intelligence are probably similar. However, men are more likely to produce super-geniuses like Albert Einstein and super-morons like the guys who knock off liquor stores and ignite their own flatulence.[6] Women, on the other hand, are more likely to produce fewer Einsteins and fewer fart-lighters. Incredibly smart *and* incredibly dumb people are both more likely to be men. That is a provocative and intriguing argument, but it in no way deserves to be labeled "sexist," as progressive critics have done. In fact, the argument's logical extension—that a randomly selected stupid person is more likely to be a man—should be far more offensive to men than to women. (But if it were indeed true, nobody should find it offensive; scientific data is morally neutral.)

Plenty of research supports both sides of the debate. Some research favors male intelligence, other research favors females, and still other research shows no intelligence quotient (IQ) difference. But, really, whether or not a gender difference in IQ exists is beside the point. The important thing to recognize is that a gender difference in IQ is *scientifically plausible* and, therefore, worthy of further research.

Consider the following: one study demonstrated that, after differences in body size were controlled for, men possess larger brains than women by about 100 grams.[7] The authors claimed that the resulting difference in intelligence was only 3.6 points and therefore did not have a practical impact on day-to-day behavior. However, this small difference in IQ could translate into an enormous difference at the high end of the intelligence distribution where men possibly outnumber women by a considerable amount. Another study suggested that men and women have the same general intelligence but manifest it differently because of anatomical differences in brain structure.[8] Specifically, men have more "gray matter" (which is related to information processing, such as in mathematics), while women have more "white matter" (which is related to information integration, such as in language). Still, other research suggested that girls' grades in math are as good as or better than those of boys, but as adults women avoid math-intensive fields.[9] (Maybe the "Math class is tough" Barbie doll is partially to blame for that.)[10]

So the jury is still out. It probably will continue to be in doubt for the rest of human existence. Obviously, comparing intelligence between genders makes people uncomfortable. Indeed, the idea that genetics plays any role at all in intelligence makes people uncomfortable, yet it is probably true to at least some extent. In an article aptly titled "Balls and Brains," the *Economist* described a study that demonstrated a link between intelligence and the quality of a man's sperm.[11] The correlation persisted even after factors (e.g., smoking and obesity) that may have confounded the results were controlled for, suggesting that genetically healthy men are smarter and have better sperm.

Even more taboo is discussing genetic differences between races and ethnicities. Dr. Esteban González Burchard at the University of California–San Francisco is a leading researcher in this field. He has uncovered differences between Latino ethnic groups in both asthma prevalence and response to asthma medication, neither of which can be fully explained by socioeconomic or environmental factors alone.[12] Of course, discoveries such as this lead to the discomforting possibility of ethnic differences in intelligence. What if a difference were found? Does that imply one group of people is genetically superior to another? This issue was explored in an extremely controversial book called *The Bell Curve*, in which the authors concluded that genetic factors are at least partially responsible for differences in IQ between whites and blacks. Predictably, this caused a public outrage, and the authors were accused of justifying racism. Without trying to speculate about the authors' personal feelings or motivations, we want to emphasize that the real issue once again is whether or not the claims are scientifically accurate. In this case, the answer was no. The authors made faulty assumptions and failed, in particular, to properly interpret their data in light of environmental factors.[13] The claims made by the authors are widely considered debunked.

Even though issues like the heritability of intelligence are unsettling, we should not ignore them. Yet that is essentially what progressives advocate. To them, merely noting that such outcomes are scientifically plausible or worthy of study is offensive. This is preposterous. Burying our heads in the sand when science investigates uncomfortable truths is not compatible with a rigorous pursuit of knowledge. Curious scientists have a responsibility to ask important questions, even if those questions make us uneasy. Instead of condemnation, the proper way to handle controversial claims is to present better evidence supporting the opposite viewpoint. It's just like freedom of speech. The solution to offensive speech is not censorship but better speech. Hence, the solution to bad research is smarter research.

Intelligence is just the tip of the iceberg. What other potentially troublesome differences exist between men and women?

Men Want Boobs and Women Want Money

This probably doesn't come as much of a surprise to you, but men really like sexy, young women with big boobs. Women, on the other hand, really like rich, powerful men. Is that sexist? Only if you disregard millions of years of mammalian evolution. Any evolutionary psychologist will tell you the explanation is straightforward: men want young, healthy women to carry their offspring (though you may disagree that a pair of artificial 38 DD breasts will really help in raising children). Women want men who can provide a secure household in which to raise children (though you may disagree that bringing home an enormous paycheck is the best metric for domestic security). Evolutionary psychologists argue that these desires are largely subconscious. Regardless of how accurate the field actually is, at the very least it provides a plausible explanation for why rich, older men constantly hook up with young women who are barely out of college. And it at least partially explains Hugh Hefner.

A beautiful evolutionary psychology experiment demonstrated that these insights are largely true.[14] Psychologists showed women pictures of a man sitting in a luxury car or a battered clunker. They were asked to rate the attractiveness of the man on a scale from 1 to 10. Keep in mind it is the same man in both pictures. The women, predictably, found the man more attractive when he was sitting in the luxury car. So what? Men would perceive a rich woman to be more attractive, too, right? Wrong. In the study, men could not have cared less about the car in which the woman was sitting. They rated her attractiveness completely on her personal appearance alone. In this very simple experiment, the psychologists demonstrated that wealth and status play a role in how attractive a woman found a man, but not the other way around. A more controversial study even claimed a correlation between a male partner's income and how frequently the woman experienced an orgasm during sex, suggesting that women are somehow programmed to be gold diggers.[15]

But the controversy over that study pales in comparison to what has dogged research that suggests women go through semen withdrawal if they do not have sex often enough with a man.[16] Semen contains hormones with antidepressant properties, and, according to this research, women who have sex without a condom receive an emotional boost that women who use a condom would not otherwise receive. Playing off this idea, respected professor Dr. Lazar Greenfield joked that semen is a gift better than chocolate. As a result, he was accused of sexism and resigned both as editor of *Surgery News* and as president-elect of the American College of Surgeons.[17] The progressive censorship machine claimed another victim.

Despite the eagerness of progressives to accuse their political opponents of sexism, it is obvious to most biologists that men and women are quite different. Because we look at and experience each other differently, it should not be a surprise that we treat each other differently as well. This is probably due to a combination of both biological and cultural factors. Today, at least in Western countries, men are more likely to open doors, give compliments, and offer to drive on long journeys—horrible stuff if you're a modern-day progressive.[18]

In fact, this sort of behavior is so horrible that progressives have a name for it "benevolent sexism." Consider a study published in the *Psychology of Women Quarterly* that asked male and female students to note incidents of sexism in a diary; the study suggested to participants a list of common sexist incidents they should watch for.[19] Telling a woman to stay in the kitchen would qualify, but so would opening a door for a woman. Do you believe that women should be rescued first in a disaster? You're a sexist pig. Tired of hearing about sexism? That qualified as sexism, too. Unsurprisingly, because of the arbitrary nature of the definition of "sexism," the authors concluded that not only were men sexist, but also that women were sexist *against themselves*. The study then went on to discuss possible interventions to make our society a less sexist place.

There is only one problem with this "scientific" study: it's not science. Actually, it is politically motivated, fluffy, feel-good psychobabble masquerading as science. Sage, which publishes the *Psychology of Women Quarterly*, describes the journal as "a feminist, scientific, peer-reviewed journal."[20] The word "scientific" is a problem because a real science journal shouldn't have to reassure its readers that what they are reading is science; it is instead reminiscent of hole-in-the-wall motels that reassure potential customers by advertising clean bed sheets. Not a good sign. And the word "feminist" is particularly problematic because it blatantly indicates the journal has a political or social agenda, not a scientific one. The article brazenly admitted that its goal was "to reduce endorsement of . . . Benevolent Sexist beliefs."[21] But stating a preconceived ideological goal is not allowed in science. Would anyone take a climate change journal seriously if it described itself as a "skeptical" journal? Or how about an evolution journal that described itself as "design oriented"? Of course, these journals would be correctly dismissed as agenda-driven propaganda, not science.

Further, a basic tenet of science is reproducibility. Science performed in the United States should yield the same results as science performed in Japan, in Russia, or on the planet Mars. Yet that is incredibly unlikely in this study because the definition of sexism appeared almost arbitrary. According to the authors, sexism was any differential treatment based on gender, even a positive treatment. This definition of sexism is simply not accepted in all societies and probably not even by most Americans. If the experiment were repeated in France, would we get the same result? How about in Asia? Or the Middle East? Although a social science like anthropology can draw comparisons between the treatment of gender in such environments, it goes against the very spirit of science to draw conclusions about all of human behavior if the underlying research does not apply to all humans, regardless of circumstance.

To summarize, an article published in an allegedly scientific journal stated, as its goal, that it intended to reduce sexism in society. Unsurprisingly, the researchers found that sexism existed and then suggested ways to eliminate it. (Tellingly, the article did not address sexism against men, indicating the possibility that political correctness—or worse, misandry—played a role in the investigation.) This sort of pseudo-intellectual propagandizing might pass as good politics, but it utterly fails as science. It should be taken about as seriously as an article written by the donut industry that concludes donuts are good for you.

The progressive insistence on absolute equality has led to denial of the biological and psychological reality that men and women are different. But, there are even more examples that may demonstrate this fundamental truth. Indeed, men and women differ in ways that scientists never dreamed possible.

Men vs. Women:
More Than Just Wee-Wees and Hoo-Hoos

A long-standing controversy about human behavior involves the so-called nature versus nurture debate. Those who espouse the "nature" side contend that genetics plays the more important role in determining human behavior, whereas those who espouse the "nurture" side—often equality-obsessed progressives and feminists—believe that social and cultural factors mostly determine our behavior. Of course, the truth is most likely some combination of the two. However, merely suggesting that genetics may influence gender roles or other gender-related traits could lead to charges of sexism.

But research from Penn State University powerfully demonstrates the influence of genetics in predisposing people toward certain types of behavior. Congenital adrenal hyperplasia (CAH) is a genetic disease that "masculinizes" girls because it exposes them to greater amounts of hormones that promote male characteristics. More severe cases result in clitoral enlargement, causing the female genitalia to resemble a penis. The researchers discovered that women afflicted with CAH were more interested in traditionally male professions, specifically jobs oriented around "things" instead of people.[22]

More importantly, the researchers discovered a "dose-response," which is vitally important in medical and epidemiological studies of causation. The concept behind dose-response is simple: the more "dose" given (e.g., a hormone), the bigger the expected response (e.g., a behavior). For instance, the more garlic you eat, the more you smell like a goat. Or as the Notorious B.I.G. would warn us, the more money you make, the more problems you have. In this study, women exposed to the most male hormones showed the greatest interest in traditionally male careers. One of the authors commented, "Maybe women aren't going into STEM [science, technology, engineering, math] careers because what they're interested in—people—isn't consistent with an interest in STEM careers."[23] Thus, the supposed "gender gap" in science, which we explore further in Chapter 14, may be at least partially influenced by our hormones.

Related to this is our ability to tolerate risk. Risk-taking is a behavior that depends on both environmental and genetic factors. Children growing up in the home of a stunt driver might be expected to display more risk-taking themselves. But there is also a strong genetic component linked to the male hormone testosterone. One study in the budding field of neuroeconomics, which attempts to explain economic behavior in terms of neurobiology and psychology, demonstrated that people who have higher levels of testosterone are likelier to engage in risky financial behavior.[24] Perhaps this partially explains why men are more likely to have a gambling addiction or a career on Wall Street. (Some people believe those are actually the same thing.) Even on the game show *Jeopardy!*, men were not influenced by the gender of their opponent when they gambled for money, but women were much more conservative if their opponents were men.[25] This suggests both genetic and social factors play a role in the crazy intersection among risk-taking, finance, and gender relations.

Progressives would prefer to cut science out of this question. They would rather focus exclusively on social factors and gender roles when looking at the development of human behavior. They prefer to blame sexist men or an oppressive society for forcing women to behave differently. Many of them refuse to accept the increasingly clear fact that genetics enormously influences our behavior, including our personal preferences. A common notion among progressives today is that girls like the color pink because our society tells them to like the color pink. After all, the color pink surrounds baby girls, whereas the color blue surrounds baby boys. According to them, gender-insensitive parents have hoisted this color preference upon all females in society.

Yet science has evidence to the contrary. Researchers showed that men and women, regardless of culture, had different color preferences. In both the United States and China, men preferred blue-green hues, whereas women preferred red hues.[26] What about Barbie dolls? Surely, girls prefer them because parents make them play with dolls. Even here, science suggests genetics plays a role. Biologists studying chimpanzees observed that even though both young males and females used sticks as toys, the females were more likely to treat them as dolls.[27] Other research has suggested that women possess a genetic predisposition to be nicer and more charitable than men.[28]

Men and women also tend to react very differently in stressful situations. While under stress and shown pictures of angry faces, men tended to disengage from those faces, whereas women engaged more.[29] Whether due to biology or social factors, the experiment suggested that women are more empathetic toward others. Bizarrely, researchers have even demonstrated fundamental immunological differences between the genders. Scientists examining gene activation patterns in immune cells in men and women who suffered from post-traumatic stress disorder made an unexpected discovery: they found that women, but not men, exhibited increased immunological activation.[30] Obviously, any differences in cellular biology are almost certainly due to biological rather than social factors.

In further support of that finding, microbiology researchers told us that male and female mice must be segregated during experiments because they have observed differences in their immunological responses. It gets weirder. Rats show gender differences in addictive behavior if they are given cocaine. Specifically, female rats were likelier than males to choose cocaine over food.[31] So if your house has a rat infestation problem, now you know that cocaine might be more tempting as bait than cheese—at least for the female rats.

Multiple lines of investigation—from psychology and animal behavior to genetics and immunology—reaffirm the basic biological fact that males and females are inherently different, often in very profound ways. Of course, social and cultural factors contribute to gender differences as well. But many modern-day progressives cling to the notion that most gender differences are the result of an oppressive, misogynistic society, and they reject the overwhelming evidence uncovered by biology. Indeed, still prevalent in progressive circles is the famous belief of feminist icon Simone de Beauvoir that "one is not born but becomes a woman." In other words, women behave the way they do not because of biology, but because of society.[32]

There is *some* truth to that statement. For instance, bragging about one's sexual conquests is culturally acceptable for men, but completely taboo for women. However, to claim that all or even most gender differences are due to social constructs is absurd. Scientifically testable biology has painted a completely different version of reality than the one created by feminist armchair philosophers. In fact, their beliefs are so contradictory to mainstream biology that it is fair to compare their anti-science rhetoric to that of evolution-denying creationists. Yet even though it is common to hear religious conservatives labeled "anti-science," progressive feminists get a free pass. Worse, if you dare to disagree with them, they will attempt to ruin your career. Just ask Lawrence Summers.

This nonsense must come to an end for science to maintain its focus on excellence and data instead of cultural whims. Even though sexism and discrimination have not been entirely eliminated from society, it is scientifically inaccurate to attribute all or most gender differences to socially contrived gender roles.

Now, having put that issue to rest and definitively restored science to its proper glory when it comes to gender, we want to turn our attention to one final question. Almost everyone on earth has debated it. This final question is of such deep philosophical significance that probably every person in the world has a strong, unwavering opinion on the matter: who are better drivers, men or women?

One study considered the fact that men spend more time behind the wheel than women do. When factoring this into their calculation, the researchers determined that women cause more traffic accidents than men.[33] Some data from neurobiology and psychology studies would support this conclusion. For instance, boys are better than girls at spatial thinking, and this ability can be detected in boys as young as five months old.[34] Perhaps men are less likely to get into car accidents because they have innately sophisticated spatial-temporal reasoning. This could also explain why there are more men than women in some areas of science academia, as we discussed earlier.

On the other hand, men are more likely to be involved in fatal car accidents, presumably because they take more risks behind the wheel than women do.[35] Besides increasing the likelihood of racing across railroad tracks to beat an oncoming train, the testosterone that surges through the bodies of men causes other problems as well. On Interstate 5 near Seattle, a multi-vehicle car accident was made worse when other (probably male) drivers crashed their cars while ogling at girls wearing shorty-shorts on the side of the road.[36]

Sadly, science will not settle the driving debate anytime soon. For now, we will have to remain satisfied with criticizing each other's driving while taking refuge in the relative safety of the backseat.

Chapter Eleven

EDUCATION AND ITS DISCONTENTS

It Stinks Being Asian in a Progressive World

By the time they have to start applying to college, savvy students are aware that colleges are looking closely at their ethnicity. The trouble starts if their ethnicity happens to be one that colleges aren't as eager to attract in an undefined diversity mixture. Asian students in particular have begun to face a peculiar side effect of our diversity-conscious education system: though a demographic minority in this country, they face an uphill battle actually getting into college.

As of the 2010 census, there are almost 15 million Asians in the United States, 4.8 percent of the population.[1] In colleges and universities that are racially blind, they are overrepresented in science and math departments. California Institute of Technology (Caltech), for example, does not use ethnicity in admissions and is 39 percent Asian.[2]

Yet it's a different story at schools that consider race, and it's been that way for some time. Thomas J. Espenshade is professor of sociology and faculty associate of the Office of Population Research and director of the National Study of College Experience at Princeton. In *No Longer Separate, Not Yet Equal: Race and Class in Elite College Admission and Campus Life*, he and

Alexandria Walton Radford looked at college applications from 1997, when the maximum SAT score was 1600. They found that African American students with an SAT of 1100 had an equal chance of getting into an elite school as a white student with an SAT score of 1410—but Asian Americans, a much smaller minority in the United States, needed a 1550 SAT score to have the same chance.[3]

Students are acting on this information. An Associated Press investigation delved into cases where Asian students with a Caucasian last name self-identified as white to improve their chances of getting into their school of choice—and other Asians are the ones encouraging them to deny being Asian.[4]

This reverse racism has become a systemic defect in our education system. Austin, which has been jokingly called "San Francisco, Texas," by residents because of its progressive bent, is also the home of the University of Texas–Austin, which has been sued for racism by a white female. Abigail Noel Fisher, the student making the bias charge, was not in the top 10 percent of her high school graduating class. As a result, she was not automatically accepted into a U of T program, as would otherwise have been the case. She alleged discrimination, suggesting that the University of Texas used racial profiling as a criterion for accepting people outside of automatic admission. For its part, the University of Texas didn't dispute that race was a factor.[5]

University representatives claim that using ethnicity as a factor in admissions is all in the interest of promoting diversity among the student body. Critics contend the university is instead penalizing students based on skin color. The outcome of the case will likely have been decided by the Supreme Court by the time of this book's release. Its ramifications may have an enormous effect on the system. In the meantime, Fisher went to Louisiana State University instead.

We are not here to argue the pros and cons of affirmative action or racial quotas or the underlying motivations and history behind them. As throughout the rest of this book, we are looking at the skewed outcomes of a world that puts ideological values above excellence. In the education system that we now confront, only truly rebellious students would dare to admit to their Asian heritage, for demographics have become the enemy of meritocracy. The fact is that specific ideas about how our society should function have infused our discussion of education, and students are not the only victims.

Science, which requires the best and brightest, is clearly at risk when schools put social engineering ahead of ability. So let us begin by surveying the landscape of American education, especially when it comes to science.

The Frames Have It

The need for more science education is so ingrained in the science lexicon that it has practically become a mantra. This makes some sense; the world is becoming increasingly competitive, so science education will remain important to any country that wants to claim or retain science leadership.

Where does American education stand? To listen to some critics, American education is terrible, dismal, caught in a downward spiral, and almost beyond recovery. "Even the most advanced U.S. students perform at or near the bottom on international assessments," claims the National Science Foundation (NSF).[6] "Our more advanced students tested in mathematics and physics were well below the international average," wrote the National Center for Education Statistics of the U.S. Department of Education in 1998,[7] while the Defense Advanced Research Projects Agency (DARPA) contends that American students and their lack of science knowledge are a national security risk waiting to happen: "Finding the right people with increasingly specialized talent is becoming more difficult and will continue to add risk to a wide range of DoD systems that include software development."[8]

Yet when it comes to results, American science is #1 in the world. The best and brightest from all over move here to learn, and that is why NSF data shows that foreign students account for more than 36 percent of doctorate degrees in the sciences.[9] Furthermore, America's scientists, concentrated mainly in academia, remain a powerhouse when it comes to delivering results. America, which has less than 5 percent of the world's population, produces almost 32 percent of the world's science.[10]

Clearly, we have a problem when we can't even agree on the nature of the problem. For how can science education be so bad that President Obama said, "The quality of our math and science education lags behind many other nations,"[11] while at the same time American science functions so effectively that it has been the runaway leader in Nobel prizes and peer-reviewed research for most of the past century?[12]

The answer is framing, which is essentially creating a thought structure so that people understand the parts you want them to understand. Political pundits, including those masking as science writers, love framing. In its idealized form, the pundits and scientists who engage in it simply believe science issues are too complex for most people to understand, so they set out to "frame" the parts they want the public to know, as you would crop a picture.

This isn't a secret strategy. Berkeley cognitive science professor George Lakoff has devoted whole books to explaining how framing can mean the difference between swinging policy in a progressive direction and not doing so. In his book *Don't Think of an Elephant*, he recalled with frustrated admiration how effective a re-framer George W. Bush was:

> On the day that George W. Bush arrived in the White House, the phrase *tax* relief started coming out of the White House. . . . Think of the framing for *relief*. For there to be relief there must be an affliction, an afflicted party, and a reliever who removes the affliction and is therefore a hero. And if people try to stop the hero, those people are villains for trying to prevent relief. . . . The language that evokes the frame comes out of the White House, and it goes into press releases, goes to every radio station, every TV station, every newspaper. . . . And soon the Democrats are using *tax relief*—and shooting themselves in the foot.[13]

As excellent a professor and scientist as he may have been, Lakoff wasn't actually interested in stopping framing so much as re-purposing the strategy for progressive aims. Forget about the truth. In his words, "The facts themselves won't set you free; you have to frame facts properly before they have the meaning you want them to convey."[14] We respectfully disagree. Coming as we do from a science background, we believe that facts need to be shorn of their frames if anyone's going to be set free.

To judge from the raw statistics, there's plenty of cause for optimism about the prospects for America's continued leadership in science, but progressives try to frame it otherwise. Some things just don't frame very well, like the Sistine Chapel and science education. If you frame the Sistine Chapel selectively enough, you might end up with just a naked guy pointing a finger—and that is a pretty good analogy for framing science education issues.

How do progressives frame science education? They insist we are actually not very good at it, and if you accept their closed system that America is not very good at science education, you can also accept that the reason is not enough money.

In actuality, science understanding among adults has increased sharply in the last few decades, almost tripling since 1988.[15] As for kids, since No Child Left Behind (NCLB) was signed into law on January 8, 2002, by President George W. Bush, test scores for students have risen nationwide. Female students have parity with males in mathematics for the first time in history.[16] Prior to this jump in the stats, progressive complaints about the crisis of education could have been considered accurate—everyone knew it and that is why NCLB passed 91–8 in the Senate and 384–41 in the House. In a rare display of bipartisanship, Republican John Boehner and Democrat Ted Kennedy were on the same side, agreeing that education was broken and needed to be fixed fast. NCLB implemented standards-based education and required states that got federal funding to issue statewide standardized tests. And it worked. Since then, things went on the upswing—until the pundits and their framing kicked in.

How Did Education Get So Bad?

From the moment of its passage into law in 2002, NCLB became the target of an all-out assault by progressives. There is a reason progressives reacted so harshly to the changes. The modern bureaucracy of education that NCLB set out to reform was created by progressives in the early part of the twentieth century—and they wanted to keep it.

Long left unmanaged, American education was in dire straits by the end of the nineteenth century. Well-meaning reformers from the newly energized progressive movement stepped in to help. Initially, the theories and strategies they designed were intended to combat inequity in learning. It was only later that progressive ideals became self-contradictory and created inequity. In the mid-1800s, Friedrich Froebel introduced the idea of educating the "whole child," in which learning would involve more than just subject matter and would focus on the interests of the child. John Dewey, the father of progressive education, seized on that idea.

But rather than keep what worked and improve what did not, Dewey set out to reshape education from the ground up. His concept of the tiered school system, for example, was not created to improve learning; it was actually designed for social acclimation reasons based on the latest pop psychology. From Dewey we also got lasting union control of the learning industry; the enduring legacy of the male superintendent who supervises employees, most of whom are women; and an elected school board consisting of elites making pedagogical decisions.

Progressives complained that NCLB's provisions about standards led to "teaching to the test" rather than to teaching critical thinking. Teaching methodology over facts has a key place in the progressive mind set and was the brainchild of a Dewey contemporary named E. L. Thorndike. Thorndike had created a stimulus-response definition of intelligence in the 1920s after surveying 8,000 high school students.[17] He imagined the human mind as a switchboard with neural bonds connecting stimuli and responses. He believed that students of higher intelligence formed those bonds more quickly than students of lower intellect, and by focusing on thinking rather than teaching facts, the schools would be more easily able to identify the students who matched his idea of intelligence. It was the perfect complement to Dewey's beliefs. With a plan for how elites could stand out from other students, Readin', Ritin', and 'Rithmetic gave way to word problems, and vocational programs were created for those with less psychological "connectionism" aptitude. Progressives thereby were nurturing elites *and* meeting the labor demands of America's burgeoning industrial society. To conservatives and liberals, that sort of curricular differentiation is clearly a euphemism for progressive social control. It didn't go unnoticed even then. Aldous Huxley wrote in 1932's *Brave New World*, "All conditioning aims at that: making people like their inescapable social destiny."[18] That is exactly what the new progressive education was doing. Yet if you ask people today whether students should learn facts or how to think, most will reply that children must learn how to think. Then we get complaints about how dumb American kids are because they don't know where Switzerland is on a world map.

Dewey oversaw the establishment of the modern National Education Association (NEA) and the militantly progressive standard it still has today, but he can't be blamed for the free fall of education that occurred at the end of the twentieth century. Responsibility for that has to fall with the U.S. Department of Education, established in 1979 during the Carter administration, which made the NEA the most powerful force for shaping the education of American youth until NCLB was passed.

Given its origins, you can deduce how the progressive social control mentality led to bizarre education "reform" attempts, such as outcome-based education, forced busing of minorities, and even Ebonics. By the year 2000, education badly needed to be fixed and everyone in both political parties knew it. The gap between whites and minorities was widening, the gap between males and females remained, and international students were pulling ahead. In early 2002, NCLB was born, and the results were immediate. By 2006, statistics showed that math and reading scores for Hispanics and African American children were at an all-time high, while thirteen-year-olds earned the highest math scores ever recorded.[19] Though it passed with overwhelming bipartisan support, NCLB was met with resistance by educators who believed that they were being told what to teach and that the school would lose funding if performance targets were not met. Even some advocates disliked its methods. In the words of prominent education expert Diane Ravitch (whose thinking was instrumental in conceiving NCLB), "NCLB was all sticks and no carrots. Test-based accountability—not standards—became our national education policy."[20] Other critics like Tom Loveless of the Brookings Institution outright reframed and reinterpreted the results, lamenting that American students were in the middle of the pack of advanced nations when it came to science and in the bottom third when it came to math.[21] This constituted major reframing because those results represented improvements; America had never been #1 in math, or science either, at the K–12 level. NCLB scores actually improved results.

In 1964, American thirteen-year-olds took the First International Math Study and ended up ranking in eleventh place—out of only twelve countries. Between the 2006 and 2009 Program for International Student Assessment tests, American scores increased five points in reading, thirteen points in math, and thirteen points in science. America had gone back to teaching information rather than trying to teach how to think. The improvement was noticeable. The Organization for Economic Cooperation and Development (OECD) in Paris has been issuing the standardized test that gets all of the attention since the year 2000, and it measures the math, science, and reading abilities of students in thirty-four countries. Its most recent version, in 2009, was given to 470,000 students worldwide, including 5,233 from 165 public and private schools in the United States. In science, U.S. students ranked seventeen out of thirty-four.[22] Not great, but not eleven out of twelve either.

OECD results put Chinese students at the top of the heap. It's worth asking, if NCLB's focus on testing and accountability is causing American students to flounder, then how do the Chinese teach and get such sterling results? Teachers are paid very little. (The wage in China is about $350 per month, even in expensive Beijing, with additional duties including cleaning the bathrooms.)[23] Learning is primarily rote. It's a simple system. Students are told what they are going to be taught. They are taught. To round things out, they are told what they were taught. Tests cover the material, without any progressive-esque focus on personality-forming or self-esteem-building and no framing of education through social justice issues. The curricula are centralized throughout China—in fact, the whole system resembles exactly the method that NCLB was intended to apply. The icing on the cake is that American children have scored higher on the OECD test every time they have taken it since No Child Left Behind was instituted—and in the last test, despite all the teeth-gnashing about American student scores, our kids beat France, Germany, and England.[24] It would seem that these facts pretty decisively favor not only letting NCLB run its course but also broadening it to include science, especially if rising up the OECD's ranks is the primary measure of American education.

But that didn't fit into the progressive framing on America's education needs. NCLB continues to rankle the education community around the country, positive statistics aside. Educators dislike it because they do not feel that they are teaching, states do not like it because expensive union contracts are foisted on them, and unions do not like it because, as we will see, they fight anything that resembles a meritocracy or that allows one school or teacher to seem better than another.

Certainly, there are other aspects of our education system that could use some attention. Teaching remains a difficult and poorly paying job. The teachers union is enormously powerful in influencing local, state, and national policy. The union's seniority rules tend to insulate burnt-out teachers at the expense of young go-getters. Whenever state budgets get tight, the first jobs on the chopping block are the young idealists, whose training is freshest, whose enthusiasm is highest, and who could be doing the most good. Microsoft Corp. chairman Bill Gates, whose Seattle-based Bill and Melinda Gates Foundation funds education programs, said in 2009 that the results of the recent test were clear: teacher pay should be based on merit to ensure that the best people are teaching.[25] That was the case in other countries ahead of the United States, according to the OECD, and it just made sense.[26] President Barack Obama and Education Secretary Arne Duncan agreed and called for a revised national education policy stressing performance rather than simply seniority, as union rules dictate.[27] People whose job it is to add more teachers, like Francis Eberle, the executive director of the National Science Teachers Association, disagreed, insisting the solution is to add more educators across the board and leave the rules and unions as they are.[28]

Yet here the party-line claims deserve a bit more scrutiny— we would benefit from looking back to Lakoff's rules for reframing. When someone like Eberle says we need more money for teachers and science outreach, it sounds convincing enough. Why wouldn't we want to get more funding and more educators? But the framing is selective: when he suggests that we "need" more money and more people working, he intentionally makes it sound like we have a deficit. Facts say otherwise; throwing money and numbers at a problem that doesn't exist is no way to pursue educational goals. The problem that unquestionably *does* exist is our antiquated, even corrupt, approach to paying and retaining the longest-serving teachers instead of the highest-quality teachers. In times of tight budgets, modifying policies to allow the best teachers to flourish and to improve students along with them seems both the smartest and the most cost-effective way to improve science education in this country.

We Don't Need More Teachers; We Need to Keep the Ones We Have

When progressives can't hijack an issue, like calling genetic modification "Frankenfood," they declare an all-out culture war. Though their self-proclaimed imagery is a peaceful dove, when it comes to social totalitarianism, they are as hawkish as it gets. It's a crucial part of their history.

H. G. Wells, the legendary science fiction author, was also founder of the Progressive League in Britain. He made no secret of his belief that progressive elites were the wave of the future, and like today's progressives he wrapped himself in a flag of loving science to rationalize the belief that progressive leadership should come first. He wrote in *A Modern Utopia*, "Science stands, a too competent servant, behind her wrangling underbred masters, holding out resources, devices, and remedies they are too stupid to use." If that sentiment seems familiar today, you read a lot of mainstream science journalism. In *The Open Conspiracy*, he was even more overt, calling for "Liberal Fascism," a war to reconstruct society toward progressive goals.[29]

Why do progressives have such a disconnect between ideology and results when it comes to education? Why twist the statistics, wage war on common sense, and desperately seek to fix what ain't broken? Grasping why science education needs to be constantly fixed by progressives even when it is doing relatively well—as we said, with adult science literacy tripled and test scores up for males *and* females (the group most often left behind before NCLB)—requires an understanding of the difference between real problems and manufactured ones.

The manufactured problem, whether as a result of framing or of simple disregard of the facts, is that America is experiencing a deficit in science, technology, engineering, and math education. Students are falling behind in these areas, advocates claim, and the downward slide can be fixed only by funneling more taxpayer money into outreach and hiring more teachers.

But let's look at actual figures, for example, fiscal year 2010. That year, the federal government funded ninety-nine different programs involved in promoting STEM education at a cost of $4.76 billion. There was hardly a lack of effort to increase STEM participation and no evidence more funding makes the difference.[30] Cuba, Namibia, Belarus, and thirty-three other countries all spend a higher percentage of their gross domestic product on education than we do, but they don't lead the world in science. Few would disagree that, STEM concerns aside, a real problem—if not *the* real problem—is securing the best science education for each tax dollar spent. And that has nothing at all to do with hiring more teachers. More bodies and smaller classes do not ensure a better science education. Arne Duncan actually endorsed *larger* class sizes to permit schools to pay better teachers more money. Adding more teachers wasn't Duncan's priority, and on this issue he was right. Yet six months after Duncan and Obama endorsed merit pay for the best teachers, NCLB was gutted by the Obama administration and merit pay was taken off the table. The action made no sense—except that Obama was heading into an election year and needed union support.

The whole debate is topsy-turvy. Articles and books by progressives about U.S. students and the sciences use terms like "dismal" and "falling behind." But scores are actually up. And science knowledge is greater among the current generation of youth than it was in previous ones—and the generations that came before are the ones that have led and continue to lead the world in science.

Nevertheless, if there is no problem, one has to be manufactured or people in the business of *fixing* problems will go out of business. That's why the STEM deficit remains such a pernicious myth. It has reached all the way to the office of the presidency—in both Republican and Democratic administrations. In 2011, President Obama, desiring to show he cared even more about education than his predecessor, topped President George W. Bush's commitment to "bring 30,000 math and science professionals to teach in classrooms" by promising "100,000 STEM teachers over the next decade."[31] It was a ridiculous promise, and it reversed his stance following the wreckage of the 2008 financial crisis when he had embraced merit pay as a more sensible alternative to adding a massive influx of new teachers.[32] In contrast to grand promises, Education Secretary Duncan tends to understand the issue better. He has called education "the civil rights issue of our generation,"[33] and he told Bloomberg journalist John Hechinger, "Great teachers and great principals elevate the entire profession."[34] Quality over quantity is the rule.

Sure enough, the actual data does not show, when it comes to the number of teachers currently employed, that this country is experiencing any kind of problem that can be solved by hiring more teachers. Richard Ingersoll and David Perda from the University of Pennsylvania tracked teacher employment over a period of twenty years and found just the opposite—there are plenty of teachers being hired. During the period of their analysis, from 1988 to 2008, they found the supply of math and science teachers *outpaced* student enrollment in schools.[35]

Their findings were not welcomed by education progressives whose single-minded intent seems to be increasing the number of teaching jobs at any cost. "When I started this work I assumed, like everybody else, that we have a critical shortage. And it was only slowly that I came to these contrarian views. Now I'm getting hate mail from people saying that I'm undermining their arguments to politicians and college presidents about the need to train more STEM teachers," Ingersoll wrote.[36]

This is the crux of science education as an issue in American life. It is not a matter of promoting excellence; it is a matter of pursuing political priorities. To progressives, the focus is not on providing a quality science education for students. Their priority is to hire more teachers and therefore create more progressives.

With 500,000 science and math teachers in this country,[37] and statistics heading in the right direction, how can anyone claim there are not enough teachers? They do so by framing the argument in deceptive ways. They will place the focus on hundreds of thousands of job "losses" but not note, as Ingersoll and Perda show, that "the total number of K–12 students increased by 19% during the past two decades" while "the teaching force increased at over twice the rate—by 48%—of overall student enrollments."[38] They instead focus on schools where there is a high teacher turnover rate and then extrapolate that problem out to all of science education. The highest turnover rates are primarily in poor schools. But the core issue in science education, teacher retention, isn't related to neighborhood income; it's a function of teacher job satisfaction, something adding more teachers cannot solve.

Why is it so hard to retain good teachers? To start with, teachers, like anyone else, prefer to work close to where they live and be a part of the community they teach in. But if you are a young science teacher, there's no way to guarantee you'll be able to teach close to where you live—in fact, you'll likely have to look high and low to find a school with a job opening. (Never mind that there are supposedly so few teachers.) Once you've got your first posting, you are a new teacher, at the very bottom of the seniority ladder—and you're forced by union bylaws to pay teachers union dues even though the teachers union actively lobbies for senior teachers to keep their jobs at the expense of younger, newer hires like yourself.

A new science teacher thus begins his career with almost no control over his workplace, his pay, and the classes he is assigned. New teachers are usually saddled with classes filled with disruptive students and forced to teach classes that leave very little room for actual creativity. This is not what people go into teaching to do.

The data shows that retention of quality, young teachers, not the hiring of more teachers, is the solution.[39] In 2006, after President Bush declared America would hire 30,000 new science and math teachers, 26,000 quit—and the reasons had nothing to do with school funding or numbers of teachers. The system is rigged against them. An excellent new teacher full of enthusiasm has no chance at all to compete for any job except a job a teacher with three more years' experience does not want—qualified or not. When a teaching job in a better location becomes available, the teacher either takes it or, disheartened by the experience, leaves education entirely. Once again, the issue is not hiring but *retention*. And the solution to retention is rewarding better employees more—something that the teachers unions will never permit.

From a progressive vantage point, the core issues in science education are fairness and equality—not excellence. This mentality is the very antithesis of science and science education, where only the best research and the soundest conclusions flourish.

Progressive Fifth Columnists
Also Want to Set Back University Science

So far we've been contending with the issue of progressive influence over grade and high school education. The picture we've painted has been of a debate deceptively framed and policy skewed by misplaced priorities. We might hope that once students reach the university level, union influence and progressive political agendas disappear. Unfortunately, this isn't the case at all.

Since 2008, the American Federation of Teachers (AFT) has launched a full-scale marketing campaign to increase membership by drawing in new members at the university level.[40] Union support makes sense to some academics who aren't making much money—a new PhD who works as a post-doctoral fellow gets a starting salary far below what peers in the private sector make. Market forces conspire to drive academic pay downward: so long as we produce six times more PhDs each year than there are academic jobs, and with laboratories funded by taxpayers, salary dollars are inevitably doled out stingily.[41] Meanwhile, the AFT's efforts seem to be working: today, 440,000 university people are members of collective bargaining groups.

For a case study in the pernicious results of a union grip on education policy, take a look at Stony Brook University. Stony Brook is an excellent research school on Long Island, 60 miles east of New York City. It is part of the State University of New York system and ranked among the top one hundred universities and top fifty public schools by *U.S. News & World Report*[42] and among the top twenty-five "best values in public colleges and universities," according to Kiplinger.[43] Tuition there is a terrific value at $6,000 per year, but because the university's reputation and education are superior to those of most other schools in nearby states (and many nationwide) administrators would like to charge more money, something equivalent to what other comparable state schools charge. In opposition? The United University Professions union.[44] The union's opinion is that allowing superior schools to charge more would mean that not all schools are equal; more importantly, it would undermine union solidarity. In other words, the union is advocating a mandate of artificial equality—with no mention at all of the possible benefits for the students.[45]

If science and science education are to remain committed to excellence, they can't be hindered by the kind of artificial equality progressives seek to institute. Most researchers don't want to be part of a herd. Independence is what has made American science historically great, and losing that competitive streak would be a blow to future generations.

Chapter Twelve

THE DEATH OF SCIENCE JOURNALISM

And How to Resurrect It

So far, we have discussed kooky progressive beliefs about science, particularly food, nature, energy, and medicine. While conservatives are routinely criticized for their anti-science beliefs in regard to evolution and climate change, the same scrutiny is not directed at progressives. The obvious question, then, is, why haven't journalists done their job? Why does the right get crucified, while the left gets a free pass for its anti-science quackery?

Perhaps more generally, it should be noted that science journalism as a whole has been in decline the past several years. This has occurred despite the fact that 65 million people consider themselves scientifically curious.[1] Why? We believe there are two answers to these questions. First, journalism in general has shifted its focus from telling a news story toward opinion and even political activism, and this has hurt the credibility of the field. Second, science journalism specifically has been overrun by partisan interests who do not love science as much as they hate their political opponents. Prominent journalists and bloggers have used their platform to criticize conservatives, religion, and corporate malfeasance while turning a blind eye toward their political allies. Their ridicule of science denial has been unidirectional, and in so doing, they have helped substantiate the widespread notion that the media are biased against half the American public.

What Went Wrong in the Fourth Estate?

When it comes to scientific issues and policy, the public craves context and explanation. People want trusted guides to help walk them through incredibly complex issues. In other words, they want journalists savvy enough to explain difficult concepts to both voters *and* policy makers, thus shaping the conversation about the most important issues of the day.

You would think that in such an environment, science journalists would be in greater demand. Yet by any objective metric, science journalism is dying—the National Association of Science Writers says science sections of newspapers in 2005 were a third of what they had been in 1989.[2]

What gives? A few factors are important to consider. Unlike sports or politics or even economics, opinion does not count a lot in science. You can become an "expert" in political journalism circles just by becoming popular. In science, you will become popular only if you know your stuff, which is hard and takes a lot of time. This partially explains why a large portion of science journalism revolves around climate change. Yes, it's an important issue, but it's not hard to understand. Science journalists, who often have absolutely no training in science, focus on the issue because it's relatively easy to write about. The earth warms up, icebergs melt, and polar bear penises shrink.[3] (Okay, that last one is actually due to pollution, not global warming.) The point is that climate change is very visual and hence relatively easy to write about compared to, say, neutrino flavor oscillation and the Reimann hypothesis, neither of which gets a lot of press coverage.

Here's another example: according to Ben Goldacre, 80 percent of the vaccine-autism debacle was covered by general journalists instead of specialized science writers. No wonder the coverage was so bad. Equal weight was often given to both sides of the debate, even when the anti-vaccine side had no science to support it.[4]

To be fair, science journalists (and general journalists who cover science) are in a tough spot: they are expected to have expert knowledge in multiple fields, along with the ability to write about those fields creatively. That has not worked out well in recent times, as we'll explain shortly.

Meanwhile, science journalists themselves are pleading for more space in media and more public mindshare. David Rowan of *Wired UK* says science journalists are "needed today more than ever" even though no one wants to read them.[5] But Goldacre, book author and writer of the "Bad Science" column in *The Guardian*, disagrees: "You're in very big trouble when academics and other bloggers can do it better themselves. I think the mainstream has talked itself out of a role in popular science, except for wacky dumbed down stories about miracle vegetables. It won't be missed."[6]

So while science journalists insist the public needs them, the public and scientists do not agree. A good detective knows that when a crime has been committed, you interview the victim, not the criminal. Asking journalists what went wrong is of no value. Instead, let's look at what the victims—scientists and the public—have to say about the death of science journalism.

In 2010, at an American Association for the Advancement of Science symposium called "Facing the Uncertain Future of International Science Journalism," BBC News science correspondent Pallab Ghosh said he had witnessed the demise of science journalism eight years earlier.[7] At that time, claims were emerging that oil companies were funding people to go to awareness and policy conferences to instill doubt about climate change. Science journalists were horrified, and in order to squelch this strategy, they collectively adopted an activist mentality against anyone who doubted climate change. They believed that they were banding together to support objective science. Yet they had inadvertently thrown away their own objectivity. They were now cheerleaders rather than trusted guides for their readers.

By 2007, this strategy, designed to support objective truth, had led to a strikingly unjournalistic and unobjective consensus. The Royal Society in England wrote directly to sympathetic science journalists and asked them not to cover opposing claims about climate change. It was false balance to even mention skeptics, they said; the science was settled. Fortunately, not all of the journalists had given up the ghost. Neil Collins of the *Telegraph*, among others, was enraged at the condescension and collusion implicit in asking journalists not to be journalists.[8] He was alarmed that the Royal Society had been able to find only 928 papers on climate change, all of which reached an identical conclusion. No paper was cited that disputed the conclusions even slightly. Yet progressive science journalists did as was asked and reported the issue was settled. These supposed journalists did no research and showed no skepticism when it came to a pet cause. (Just to clarify: we believe that global warming is quite real. However, it is a journalist's *job* to be skeptical, not to just regurgitate whatever a scientist tells him.)

Things got worse. The 2007 UN IPCC report on climate change had been preceded by talking points for half a year; these were published by science journalists in the mainstream media as facts. Acting more like stenographers than journalists, writers dutifully reprinted the media talking points without verifying them. One example was a story from 1999 that claimed Himalayan glaciers would melt by 2035.[9] On the surface the story sounded convincing enough to science journalists—however, it completely lacked evidence. The only evidence was a consumer media article containing speculation by Dr. Syed Hasnain.[10] Corporate science journalists didn't bother to do any journalism (though it is safe to guess they would have if Exxon claimed glaciers would be growing by 2035). What had been published as fact by the media worldwide was speculation based on an unpublished paper. The lesson, according to Ghosh, was this: "Do not defend science," he cautioned the journalists in attendance. "Ask the awkward questions." What did he mean? Let's illustrate with a couple of examples.

Sports journalists do not feel a need to only discuss homeruns for baseball players. They respect the audience enough to know that readers understand the sport, and they feel free to describe the ups and downs and the drama that occurs during the season. They do not defend baseball because it doesn't need defending. Similarly, political journalists don't mind asking awkward questions. Have you ever heard a political press conference? ("Mr. Congressman, was that a photograph of your genitalia?") Journalists don't feel a need to defend politicians. They report the news—both the good and the bad—because that is their job.

However, instead of discussing science, science journalists began to think of themselves as science's defenders. Such defense was never necessary, and the idea that climate science or human embryonic stem cell research or genetic modification is so important that it needs a bunch of non-scientists to defend it from the public is both presumptuous and silly. A journalist should be a nag, piecing together the story behind the story, no matter what he or she personally believes about it. When journalists forget that, they lose their objectivity and thus their very right to call themselves journalists. They instead become stenographers or, even worse, cheerleaders. Even if they think they "love science," they are now mere steps away from turning it into a political cause.

Necessary awkward questions are surely not being asked when science journalists serve primarily as cheerleaders, often trolling for wacky stories. People are allergic to electricity! Miracle vegetables can lengthen your life! Or a statistical wobble in the universe, reprinted from NASA press releases, could be life on other planets!

The latter issue, press releases, has become a sore spot. They do usually contain solid stories because well-informed public information officers at universities often write them. However, too many of the science journalists that remain are reduced to simply rehashing press releases, what Nick Davies referred to as "churnalism"—that is, churning out press releases instead of doing original reporting. He noted a recent study of British journalism that found that 80 percent of stories were not original and only 12 percent of stories were generated by reporters rather than press releases.[11] In response, defenders of science journalism contend that overburdened writers are forced to turn to churnalism because of greater demands in a shrinking workforce. But is that true?

It's a seductive argument—few would begrudge journalists their right to complain that they are working in less-than-ideal circumstances. But churnalism is not just a modern-day phenomenon. In the 1980s and 1990s, science journalism was far from imperiled, but in 1998 Dr. Arpad Pusztai of the Rowett Research Institute in Aberdeen began claiming that rats develop stunted growth and impaired immune systems after eating genetically altered potatoes.[12] Mainstream science journalists failed to ask the awkward questions and published Pusztai's claims as fact. When it came out that these suggestions were derived from unpublished and unreviewed (and, as it turned out later, unsubstantiated) material,[13] the resulting debacle looked a lot like the melting Himalayan glaciers in the IPCC report. It wasn't the first time science journalists let phony facts slip by because it matched their advocacy or made for good scare journalism. The Alar apple scare, cold fusion, and even—as we've seen earlier in this book—DDT in the 1960s are examples of stories in which journalists trusted press releases, pundits, and hot air over evidence-based science. All of those flubs came long before the present era of budget cuts in newsrooms.

The strangest thing about these specious flaps is that besides selling newspapers, they also led to greater amounts of money pouring into research into the topic in question. As journalist Gary Taubes said of the phenomenon that plagues both science and journalism,

> I used to joke with my friends in the physics community that if you want to cleanse your discipline of the worst scientists in it, every three or four years, you should have someone publish a bogus paper claiming to make some remarkable new discovery—infinite free energy or ESP, or something suitably cosmic like that. Then you have it published in a legitimate journal; it shows up on the front page of the *New York Times*, and within two months, every bad scientist in the field will be working on it. Then you just take the ones who publish papers claiming to replicate the effect, and you throw them out of the field.[14]

Ouch. That's not exactly a ringing endorsement of science journalism.

Progressive Activism Takes Over

Now that we've put to rest the notion that today's science journalists are uniquely crunched, we can definitively address the question "How did science journalism collapse so dramatically?" The larger trends that have led to many Americans leaving newspapers and mainstream news behind altogether have been in the works for decades.

Journalism has historically been plagued by sensationalism, but the post–World War II era was a "golden age" in America in many ways and in many professions, including journalism. The 1950's prevailing climate of racism and sexism aside, standards of living were unmatched in history, the economy was booming, and people were in a position for the first time in a few decades to broaden their knowledge. The demand for quality writing skyrocketed.

Journalism then, like music, was something you learned by doing. Young journalists learned their craft by doing the jobs senior journalists no longer had to do; they covered midnight crime scenes and night court and drug addicts and people in poverty and wrote it all up. The writers may have been liberal in political outlook (after all, what young person wouldn't think society could find a better way after witnessing all that?), but editors made journalists stick to the facts. *People are smart, so just give them the story* was the common theme.

Yet by the late 1960s a liberal democracy basking in wealth was transforming into a progressive one intent on implementing more radical change—and journalism was not immune. Walter Cronkite of *CBS News* shucked off impartiality and declared he was against the Vietnam War. It wasn't the first time that journalists had been critical of wars and peace in the past. *LIFE* magazine in 1946 had stated, "Friend and foe alike, look you accusingly in the face and tell you how bitterly they are disappointed in you as an American" and "Never has American prestige in Europe been lower"[15]—and that was in that euphoric months after the end of World War II. So opinionated journalism wasn't revolutionary, much as progressives would like to think it is.

But when Cronkite took a side, it meant something. He had also covered World War II for United Press. He knew both the good and evil that men and governments could do. He had done his best to stay objective even when covering the Allied invasion of Normandy and flying in Allied bombers on missions over Germany. Deviating from objectivity was clearly painful. It had been an agonizing decision for him, and people knew it; his struggle and his conclusion changed public opinion on the Vietnam War.

A line had clearly been crossed. Journalists saw his impact on culture—President Lyndon Johnson remarked to an aide after the Vietnam declaration, "If I've lost Cronkite, I've lost America"—and they, too, wanted to "make a difference." Journalism became less about telling the story and more about changing the game. Bob Woodward and Carl Bernstein controversially used an anonymous source in covering the Watergate scandal, and that brought down a president—the ultimate making a difference. They set the bar for being a force for progressive good works even higher.

Some in the generation that followed didn't even stop with anonymous sources; they were content to use no sources at all, and sometimes when they couldn't find a story, they invented it. Stephen Glass of the *New Republic* made up stories and created fake websites and sources as supporting evidence. Janet Cooke of the *Washington Post* won a Pulitzer Prize for "Jimmy's World," a piece about an eight-year-old heroin addict who didn't exist. In September 2004, Dan Rather of *CBS News*, heir to Walter Cronkite ("the most trusted man in America"), on the nightly news broadcast reported a story about the military record of President George W. Bush. The evidence was revealed as phony within hours, but Rather stuck by it, saying the phony evidence was less important if "the major thrust" of the story about Bush were true.[16]

Journalism had clearly lost ground among the public, but there was no reason for science journalism to have suffered the same fate. In contrast to what you might believe after reading complaints from Western media journalists, science journalism is not dying everywhere. In Asia and in many developing countries more interested in progress than progressivism, science journalism, along with journalism in general, is gaining ground. Diran Onifade, the president of the Federation of African Science Journalists and vice president of the World Federation of Science Journalists, said in a 2009 interview, "If we are talking about science journalism in the context of a world of diverse time zones, it may indeed be dusk in the western hemisphere. But in Africa, it's just the dawn of a promising day."[17] But in a place like contemporary America, where ideology and "the major thrust" of the story are more important than facts and details, journalism is transforming into something that would be unrecognizable to its earlier incarnations.

Framing: The Secret Weapon of (Progressive) Science Journalists

Progressives in science journalism have always been quick to cover issues important to them (e.g., climate change) or issues that make conservatives look bad (e.g., denial of evolution). But science stories (or anything else) that make progressives look bad? Not so much.

The anti-vaccine movement is associated with the political left. Remember, it was Robert F. Kennedy Jr. who was an early and vocal proponent. Have you noticed who dresses up in ridiculous animal costumes to protest medical research? Or who wears gas masks to protest nuclear power? Kooky progressives. Are journalists quick to label of any of them "progressive crazies"? No, of course not. They are careful to ignore the obvious political demographic of these movements. But any conservative who is opposed to evolution is immediately labeled a "religious nut." Similarly, journalists frame progressive opponents of genetic modification as being anti-corporation rather than anti-science— even though all the arguments they make against GM technology are anti-scientific.

And compare all that to how conservatives have been treated in regard to the human embryonic stem cell research controversy.

Conservatives are, once again, called anti-science, religious zealots—and then that label is applied to the entire Republican Party as well. The double standard is truly stunning.

Unfortunately, when science journalists aren't busy defending progressives, they are busy looking for stories to exaggerate and blow out of proportion. Even though science journalists once contended they were critical watchdogs who tempered the hype of university press releases, scientists know better. Professor Larry Moran of the University of Toronto, a prominent blogger and critic of shoddy science journalism, disagrees that science journalists are the trusted guides for the public they claim to be: "Over the past decade it has been the 'professional' science journalists themselves who were the gullible victims of scientific hype and PR."[18]

For instance, writing in *Nature* about coral reefs, Gene Russo described a paper that stated that "increased concentration of carbon dioxide will decrease the aragonite saturation state in the tropics by 30 percent and biogenic aragonite precipitation by 14 to 30 percent."[19] But a university press release read, "Great Barrier Reef will crumble within 50 years."[20] National Public Radio then invited the lead author, marine ecologist Joanie Kleypas, to talk about global warming causing the Great Barrier Reef to collapse, not the much less exciting decreased rate of reef calcification. Russo interviewed Kleypas about that experience, who described being caught unaware by the interview: she was stuck between contradicting her collaborators, perpetuating an unfounded conclusion, or implying a lack of knowledge about her own research by skirting the issue.

The desire to shock and alarm the audience is a common tactic not only of science journalists, but also of journalists in general. Scare journalism, like the Great Barrier Reef example, is usually not scientifically sound—just like many of the chemical and medical scare stories you hear daily. Instead, scare journalism does a disservice to the public and turns goodwilled scientists into journalism skeptics.

Science journalists contend their overall demise is more of a business or economics issue or even a general journalistic trend. They also claim that people now pick their news based on their worldview and thus pick their science the same way. Whatever reason science journalists use to explain their own demise, the excuse is the same: it's not their fault. But that "blame the public" mentality defies history. People did not stop reading newspapers because of competition from radio or television. Journalists lost people's trust by imposing their worldviews on the public, and many science journalists did the same thing.

But science journalism can be revived. And in some ways, it is being revived now.

Science Journalism: A New Hope

No one can deny that there remain outstanding science journalists, like John Tierney, Carl Zimmer, and Greg Critser, just to name a few. There are also many good news outlets that specialize in science. And a new generation of science bloggers—many of whom are scientists—is communicating complex issues to the public. Therefore, we have hope that a new era of science journalism is dawning.

Also, it is important to note that scientists *want* to like journalists. Despite the popular myth of being unsociable and irritable, the majority of scientists are pleased that people are discussing their research. They want to help the public understand why their research is important.

And science journalists love science, even if they let their politics get in the way on occasion. Many have a science background, and some love science so much they toil in obscurity writing anonymous releases for university press offices. Talk about a difficult and thankless job.

In tandem with researchers, we are confident that science journalism will not only fully revive, but also be better than ever. It just has to repair its reputation of being a battlefield for political and cultural wars.

Basically, science journalists should follow three simple rules. Don't defend science. Ask the awkward questions. Be trusted guides for complex issues.

Chapter Thirteen

FALSE EQUIVALENCE

The Faulty Logic Behind Fake Logical Arguments

It was said by the French General Napoleon Bonaparte that an army marches on its stomach. His sentiment is still true in modern times, but the progressive army of today needs different provisions than just organic beans and free-range rice; it also feeds off seductive, but fallacious, logical arguments.

Not only that, progressives size up their battlefields well in advance and come armed with misleading facts and figures. They love to win with overwhelming force, overpowering their opponents with sheer numbers of other progressives saying the same thing. "Copy and paste" is a particularly useful weapon for progressive bloggers.

So who are the foot soldiers in the progressive army? Political activists—climate-hawk bloggers, anti-religion bloggers, and think tankers. Basically, they're propagandists with a keyboard.

But like Napoleon, who lost when arrogance kept him from understanding the opposition, progressive activists can also be banished to an intellectual island. Before that can happen, however, we must understand their chosen weaponry.

If You Disagree with a
Progressive, You're Anti-Science

Progressives routinely brag that their beliefs are based on science and that those who disagree with them—in particular, conservatives—are anti-science. Some go so far as to claim that science and progressivism are one and the same. They purposefully conflate technological "progress" with political "progressivism," playfully tossing around the two words as if they were synonyms. In their world, science is an intellectual pursuit that does not have room for people with different political ideologies—especially conservatives.

The prominent progressive think tank the Center for American Progress runs a website called Science Progress, which says this about itself: "Through this work we are building piece by piece the progressive narrative of science and technology policy."[1] That's an interesting way to frame science. Did you notice that statement could be read in two different ways, depending on the usage of the word "progressive"? That double entendre is no coincidence; it is just another manifestation of their program of linking technological progress and social progressivism.

Science Progress has more to say. Note the following from an article titled "Time for Science to Reclaim Its Progressive Roots":

> The phrase "science progress" is, arguably, a bit awkward. . . . It reminds us that we are the inheritors of the Enlightenment's confidence in the possibility of improving the human condition—a possibility predicated on values of individual freedom, social equality, and democratic solidarity.[2]

Heart-warming as this sentiment may be, we must ask, is it true? How much does science have to do with social equality? Should we expect the fields of biology, astrophysics, and geology to reveal truths about social justice? And exactly which part of the scientific method employs democratic solidarity? Science is about data, not democracy.

The trick here is that if scientific progress and societal progress (as they define it) were, in actuality, the same thing, those who disagree with their political vision of societal progress would be not only "anti-progress" but also "anti-science." From this arises the progressive claim that conservatives are scientifically illiterate. The media regularly perpetuate this myth, yet never pause to consider that liberals, libertarians, and progressives are just as capable of holding anti-science beliefs.

Consider the following: surveys show that 93 percent of scientists acknowledge the necessity of animal research, 62 percent of Republicans agree, but only 48 percent of Democrats concur.[3] Building more nuclear power plants is supported by 70 percent of scientists, 62 percent of Republicans, but only 45 percent of Democrats.[4] In 2004, the National Academy of Sciences reported that genetically engineered food is safe. What does the public think? More Republicans (48 percent) than Democrats (42 percent) agree with the nation's top science organization.[5] On these three major issues, Republicans are more closely aligned with the scientific community than are the Democrats. There are many more.

This stark contrast between scientists and Democrats is trivialized by progressive pundits as a mere disagreement over morality or safety. For instance, Democrats aren't anti-science because they oppose animal research in the example above; they just have higher moral standards than everybody else. Progressives aren't "anti-science" about nuclear power; they are just "pro-safety." Progressives aren't technophobic about GM foods; they are "pro-choice" about nutrition. In other words, it is okay to be anti-science as long as you have a good *personal* reason for it. However, when conservatives use the exact same flimsy excuse to ignore evolution or anthropogenic global warming, progressives have no qualms about labeling them "anti-science."

Yet if you look at polls, progressives have little reason to celebrate their alleged intellect. While progressives routinely mock conservatives for denying evolution, a substantial portion of their political allies (i.e., Democrats) also reject it. Depending on how the question is asked, different polls can yield strikingly different results. One Gallup Poll, for instance, shows that 60 percent of Republicans reject specifically human evolution.[6] However, according to a Pew poll, when asked about evolution in general, only 39 percent of Republicans reject it—but so do 30 percent of Democrats.[7] Other data suggests that when more questions about basic science are posed, progressives do not perform as well as conservatives. The blog Audacious Epigone revealed that on a whole host of scientific issues, Republicans are more knowledgeable than Democrats and Independents.[8] (The data is slightly modified and reproduced in a table on page 213; the "winning" percentage is in bold.)

However, conservatives shouldn't start jumping with joy. *Discover* magazine science writer Razib Khan reclassified the respondents (as "liberal," "moderate," or "conservative") and reanalyzed the data. His analysis showed that conservatives and liberals are roughly equal in their knowledge of science, but that both are more knowledgeable than moderates. He offered an alternative explanation: people who vote Republican because of economic issues are more scientifically literate than conservatives.[9]

Our point here is not to determine which political ideology is dumbest. Instead, we simply wish to dispel the often-repeated lie that conservatives or Republicans have a unique tendency to hold anti-science beliefs. We reject this orthodoxy, which until now has remained largely unchallenged.

We said earlier that a conservative war on science has been dramatically exceeded by calls for Armageddon by progressives.

Scientific Issue	Dem.	Ind.	Rep.
Astrology is not scientific	64.3	55.7	**75.1**
The benefits of science exceed the harms	73.3	66.2	**78.0**
Understands the need for control groups in testing	79.8	81.4	**82.1**
The earth's core is very hot	94.2	92.6	**94.6**
Demonstrates a basic understanding of probability	87.9	90.0	**91.8**
Not all radioactivity is man-made	79.2	78.5	**85.9**
Father, not mother, determines a child's sex	72.0	74.7	**77.3**
Lasers are not made by condensing sound waves	63.4	70.9	**75.1**
Electrons are smaller than atoms	71.4	71.3	**72.8**
Antibiotics do not kill viruses	55.7	55.4	**68.5**
Continental drift has and continues to occur	90.1	**90.6**	87.9
Humans evolved from other animals	**57.6**	50.7	41.5
The earth revolves around the sun	79.2	73.9	**81.5**
It takes the earth one year to rotate around the sun	75.8	78.8	**78.9**
Respondent will eat genetically modified foods	66.1	69.4	**73.1**
The North Pole is on a sheet of ice	**67.0**	59.7	63.4
Not all man-made chemicals cause cancer when eaten	46.6	46.4	**52.6**
Radioactivity exposure doesn't necessarily cause death	67.5	67.1	**77.0**
Exposure to pesticides doesn't necessarily cause cancer	55.5	57.9	**66.8**

Why? Because once it has been demonstrated that the facts don't align with the orthodoxy progressives hold, they launch an all-out assault on logic itself: enter false equivalence.

Is It Logical to Invent Logical Fallacies?

When we met with our book publisher, we were asked what we thought would be the grenade militant progressives would try to throw at us in retaliation for our arguments in this book. We answered simultaneously, "False equivalence." It's an easy strategy because it sounds intelligent even though the meaning is just fuzzy enough to baffle listeners.

Here's a good example: in response to an article Alex wrote in *USA Today* about how both Republicans and Democrats endorse anti-science policies,[10] Paul Raeburn at Knight Science Journalism Tracker responded, "Many Republicans reject evolution and climate science, and there is no example of Democrats correspondingly rejecting a theory or an entire field of science."[11]

By now, you know that statement is obviously an untrue assertion. Maybe his opinion will change after (if?) he reads our book.

But, essentially, this is the principle of false equivalence: progressives aren't perfect, but there is no equivalence between dumb progressives and dumb conservatives. Clearly, progressives are smarter. Anti-science beliefs held by progressives don't matter; only the ones held by conservatives do. Honestly, we wish this were an over-simplification of their argument, but it's not. One academic, Ron McClamrock, a tenured professor of philosophy at SUNY-Albany, summed up the progressive sentiment quite bluntly: "Lefties are overrepresented in academia because on average, we're just f-ing smarter."[12]

Instead of asking you to replace lefties in that sentence with "men" or "whites" so that you can see how bigoted and ridiculous (yet common) that belief is, we'll tackle it scientifically. Let's backtrack all the way to Logic 101. One of the first things you learn in logic class is the difference between formal and informal fallacies. Formal fallacies are errors in logical structure. For example, "All cats are mammals" and "All dogs are mammals" are true statements. However, you cannot conclude from those statements, "All cats are dogs." That is an invalid argument due to an error in structure, and this formal logical fallacy is called "fallacy of the undistributed middle."

Then there are informal fallacies. These fallacies don't violate logical structure but still represent erroneous thinking, often in the premise of the argument. This is where you heard about *ad hominem* attacks (Bob is stupid!) or our personal favorite, *post hoc ergo propter hoc.* (Isn't that just fun to say?) It means "after this, therefore because of this." An example would be claiming that sunsets cause the moon to rise because the sun always sets right before the moon comes out.

Even if you don't make use of logical fallacies in your day-to-day life, politicians and pundits do. Politicians find *argumentum ad baculum* especially useful. It means not doing A will result in B, and since B is bad, everyone should do A. In practice, that means, "Vote for my tax increase, or we'll cut funding to hungry schoolchildren," or "Vote for my candidate, or the terrorists win."

Useful as all of these fallacies are for decoding the evening news, you might not remember false equivalence from logic class. Why? Because progressive political writers made it up. They invented the term just for conservatives. False equivalence is the ultimate manifestation of progressives' relativist worldview because the definition of the term can change at the whim of the person speaking. Essentially, what progressives mean by false equivalence is that any comparison that makes them look bad is, by definition, false. It's convenient, but intellectually vacuous.

The strategy of crying false equivalence caught fire in 2011 when progressive writers in the media leveled this charge against anyone who dared discuss the growing anti-science sentiment among people on the left. Of course, progressives grudgingly conceded that sometimes they used false equivalence incorrectly, but they insisted the real problem was that conservatives just didn't understand logic at all, so it happened on the right a lot more.[13] An obvious example of false equivalence dismissal by progressives is that when Bill Maher called Sarah Palin a "cunt" on his HBO show, it offended no one on the left who claimed to care about the feelings of women, but when Rush Limbaugh called Georgetown law student Sandra Fluke a "slut" during his radio program, the left was enraged. When conservatives noted that was a double standard for conservatives, progressives shouted "false equivalence."

From what we can understand, progressives use the term to describe a comparison that is founded on deception in one of the premises. Yet by invoking false equivalence, progressives turn themselves into perpetrators of it. Most Republicans would likely acknowledge that carbon dioxide is a greenhouse gas. But since they are unwilling to curb American CO_2 emissions until science can prove Chinese, Mexican, and Indian cars do not also cause pollution, progressives claim Republicans are denying climate change and are anti-science. (Despite the rhetoric, one poll showed 60 percent of conservatives accepted climate change. More on this later.) Fine. It is true that conservatives are not willing to embrace anthropogenic global warming to the same extent as everybody else. Yet there is a great solution to the problem of global warming that conservatives support: nuclear power. But progressives are likelier to oppose nuclear power than conservatives. Does that make progressives anti-science? No. That's false equivalence.

Writers like Joe Romm of the progressive blog Think Progress must engage in some pretzel-twisting logic in order to justify progressives' anti-science position against nuclear power and the overwhelming majority of scientists who endorse it. He claims that we should not conflate "what scientists tell pollsters with science itself."[14] Apparently, he assumes that scientists hold opinions that differ from the science itself, or that it is okay to trust scientists on climate change but not on nuclear power. Scientists are to be trusted unless they are out to kill us—which happens to correlate with scientific issues that progressives don't accept.

Progressives also use disbelief in global warming as evidence that conservatives are anti-science. Applying the standard of false equivalence against progressives, we believe this is an example of false equivalence because some conservatives object only to the specific term, not to the concept. It could be that conservatives are actually more discerning of correct scientific terminology on this issue than progressives. Many in science have abandoned the term "global warming" because it was scientifically incorrect. Now the accepted term is the more scientific "climate change," meaning climate instability. Yet progressives see no real distinction between the terms—a recent poll showed that about 87 percent of Democrats believed in both global warming and climate change. By comparison, only 44 percent of Republicans believed in global warming, while 60.2 percent accepted climate change.[15] On this particular issue, conservatives actually *do* care about the scientific meaning of words in a way that does not seem to matter to progressives.

On climate change, no question, more conservatives are clearly wrong. But by using the pretzel-twisting logic of false equivalence, progressives can justify any position they want. That is routinely how progressives manipulate data to come to the conclusion that conservatives are all anti-science. It's about framing, and progressives are master manipulators of journalistic narrative.

Interestingly, while progressives accuse others of false equivalence—remember, a completely invented logical fallacy—they engage in a very real fallacy of their own: poisoning the well. Basically, this fallacy is a type of *ad hominem* attack used to discredit everything a person has to say by ridiculing some unrelated belief or action. Paul Raeburn attempted to do that with Alex's *USA Today* piece:

> You might wonder why I took a few minutes to run Berezow down [using Google]. What I found doesn't tell me much about him, but we can at least ask whether he himself is a conservative, based on where he's published some of his writing. . . . If Berezow is a conservative, that might explain why he thought Democrats were just as bad as Republicans in their attitude toward science.[16]

According to Raeburn's logic, he is implying that nothing we say is credible because we might be conservatives. He doesn't actually know, but we *might* be. So just to be safe, nothing we say should be taken seriously. It's just not possible for any smart person to disagree with Raeburn on politics. (In case you are curious, we consider ourselves to be some mix of conservative, liberal, and libertarian depending on the issue.)

But getting back to our main point, we are going to investigate in the next few sections several highly dubious false equivalence claims.

False Equivalence Claim #1:
If Both Sides Have Anti-Science Beliefs,
Then Progressives Aren't Anti-Science

Most likely, progressives on the defensive will contend the anti-vaccine movement, the anti-GMO movement, the anti-nuclear movement, and other anti-scientific beliefs—despite extensive evidence in this book—aren't actually progressive positions. They'll argue that both ideologies have people with those beliefs, so progressives cannot be held accountable. Yet they will simultaneously argue that conservatives are anti-evolution because a few percent more of Republicans reject evolution than Democrats do.

With reference to the anti-vaccine issue, progressives like to claim that because some of their leaders criticize the anti-vaccine movement, it should not count as a progressive anti-science belief; it is simply "motivated reasoning" by emotional people when progressives do it and thus false equivalence.[17] Although progressive apologists are correct that some people on the left have distanced themselves from the anti-vaccine kooks, the logic should extend to the right as well. After all, plenty of right-wing people have criticized conservatives who reject the simple physics of greenhouse gas emissions or basic geology. Despite this, progressives continue to frame all conservatives as anti-science, as if all Republicans have to sign some sort of pledge denying that pollution is bad or accepting that the earth is 6,000 years old.

False Equivalence Claim #2: Facts and Numbers That Make Progressives Look Bad Probably Aren't Reliable

Whenever a public opinion poll shows that progressives might hold an anti-science belief, they prefer to dismiss it as inaccurate or irrelevant. A 2009 Pew poll, which we discussed above, found that left-wing people are anti-science on plenty of positions.[18] However, in the opinion of many progressive writers, it was false equivalence to point this out because it doesn't matter what the masses think, only what the "elites" or leaders think.[19] Apparently, according to progressives, elites matter the most.[20] Of course, this standard is subject to change.

When Kentucky Governor Steve Beshear, a Democrat, endorsed a creationist museum in his state, progressives stated it was false equivalence to claim he was anti-science. The reason? He just cared about jobs, they rationalized.[21] However, when a Democratic legislature in Louisiana passed a law allowing the teaching of creationism in schools, it was Republican Governor Bobby Jindal who received a lion's share of the blame.[22] The law was portrayed as "stealth creationism," which is an anti-science buzzword applied only to the right.

False Equivalence Claim #3: Progressive Anti-Science Beliefs Are Not Mainstream

The Earth Liberation Front is decried by almost everyone. That its members hold anti-science beliefs is not really important to this discussion because eco-terrorists are not part of any reasonable discourse. They are not part of the mainstream.

But what about Greenpeace and the Sierra Club? Are they not part of the progressive mainstream? Despite research by the National Academy of Sciences showing that there are no adverse effects of GM crops, these environmental organizations oppose them.[23] Joining them in opposition is the dubiously named Union of Concerned Scientists, which has a long history of peddling anti-science information. The truth is GM foods have not resulted in a single stomachache in over a decade of use, whereas documented poisonings and deaths have been attributed to the consumption of organic food. Obviously, conventionally grown food also carries the risk of disease outbreaks, but the idea that GM food poses any unique health risk is not supported by scientific data.

Instead, GM crops may feed a lot of children who might otherwise starve. Because these crops increase agricultural efficiency, they require fewer of the scary-sounding chemicals that progressives despise (particularly insecticides), and the ability of poorer people in remote areas to grow more food will result in fewer carbon emissions. It will also help make them richer, since they will be able to sell their surplus crops. The economic and environmental benefits of GMOs have proved to be considerable.[24] Yet the progressive "elite"—along with the plebeians among the ranks that are dismissed by progressive writers as inconsequential—consistently opposes genetic modification. Another example is AquAdvantage® salmon, which has been modified to grow faster and feed more efficiently than its unmodified cousins. Evidence shows that increasing consumption of fish has health benefits, and GM salmon might help reduce overfishing in the ocean. That should be a win for everyone—environmentalists, nutrition advocates, scientists, and citizens.

Yet even though this fish has been thoroughly tested for safety, it has been trapped in FDA review since 1995.[25] The FDA has confirmed that it has passed every test—but it is still not feeding anyone.[26] Why? Lobbying by progressives in Congress, the very elites we are told are not anti-science. Democratic Senator Mark Begich of Alaska invoked a favorite progressive buzzword about how evil biologists are when he said, "We don't need Frankenfish threatening our fish populations and the coastal communities that rely on them."[27]

Confronted with this uncomfortable truth, progressives drag out false equivalence once again. In response to the same *USA Today* piece by Alex we discussed earlier outlining the left's anti-science positions, progressive blogger and author Chris Mooney dismissed the issue of genetic modification as false equivalence because "it is not a mainstream position, not a significant part of the liberal agenda."[28] Do seven Democrats in the U.S. Senate not count as part of the progressive mainstream agenda?[29] Has anyone alerted Greenpeace members and other progressive environmentalists that they are not a part of the mainstream discourse?

False Equivalence Claim #4: Conservative Dumb People Are Dumber Than Progressive Dumb People

Just for a moment, let's revisit the "evolution versus creationism" dead horse, because progressives love to trot out its carcass so much. It isn't unfair to suggest that progressives are allergic to religion. If a person is religious at all—or even just believes in some higher purpose—many progressives assume that person is a religious wingnut and a creationist. Some even claim that religious belief and science are incompatible, though 27 percent of scientists have belief in God, and another 30 percent are agnostic (and hence at least open to the idea of God).[30] Thus, we agree with astrophysicist Neil deGrasse Tyson when he says,

> Consider also that in America, 40% of American scientists are religious. So this notion that . . . if you're a scientist, you're an atheist, or if you're religious, you're not a scientist, that's just empirically false.[31]

Yet progressives routinely claim otherwise. They claim that conservatives—particularly religious ones—are downright stupid. They might admit that their side has dumb people, too, but that conservatives are clearly dumber. Yet using that as a line of intellectual defense is only marginally better than Pee Wee Herman's effective insult comeback, "I know you are, but what am I?"

It's the miracle of false equivalence.

Regardless of the facts, in his latest book, *The Republican Brain*, the aforementioned Chris Mooney argues that people on the right are stupid because they can't help it. Blame biology. If they could help it, they wouldn't be conservatives. Climate scientist Roger Pielke Jr. responds that Mooney's argument sounds a lot like "eugenics."[32] Of course, if Mooney had made the same argument about, say, ethnic minorities, he would be correctly called a bigot and a racist. But since he smears only conservatives, progressives consider him a respectable authority.

False Equivalence Claim #5: What Appear to Be Progressive "Anti-Science" Beliefs Are Really Just Disagreements over Morals and Ethics

As we have shown, progressives will dismiss the schism among leftists over nuclear power as a disagreement about ethics, not about a fundamental misunderstanding of risk perception or radiobiology. Their disagreement over animal research is about morals, they claim, not a fundamental distrust of biomedical science. Their disagreement over vaccines is about caution, not an anti-science stance regarding epidemiology and immunology. Their disagreement over GMOs is about concern for the environment and distrust of corporations, not a fundamental misunderstanding of genetics.

You see, when progressives oppose science, it's because they are concerned about morals or ethics. Their opposition has nothing to do with ignorance. But when social conservatives expressed concern about brand-new human embryonic stem cell research technology in 2001, it was not because they were concerned about morality but because they were anti-science, progressives claim, and still are today.

In this way, progressives routinely trivialize, dismiss, and make excuses for anti-science beliefs on their side of the aisle, yet throw the rhetorical bomb of anti-science at all their conservative opponents. This, more than anything, is the most absurd and infuriating aspect of false equivalence. It fails to properly acknowledge the reality that anti-science beliefs exist among all political worldviews—and addressing them, even when our friends hold them, is an important thing to do. Instead, the doctrine of false equivalence encourages progressives to smugly (and incorrectly) believe that they hold the intellectual upper ground.

Indeed, false equivalence is quite ironic: framing all opposing political arguments as being false equivalence may be the best false equivalence of all.

Chapter Fourteen

THE WAR ON EXCELLENCE

*Science Advances Cannot Be Framed
Through Social Justice Issues*

BY NOW, OUR LARGER POINT has been made clear: progressivism, once the philosophy of responsible policies to achieve a better world, has become disconnected from its roots. The progressivism of today has inherited the social authoritarianism and righteousness of its antecedents. But though its proponents have held on to the label of "pro-science," their adherence to a set of ironclad feel-good fallacies leads them to advocate policies that have less to do with making progress and more to do with going backward.

Instead of embracing positive scientific advancements, today's progressives are caught in a double bind. On one hand, they maintain an unrealistic faith in magical solutions. On the other hand, they squelch progress and technology by positing the impossibly burdensome precautionary principle as the standard that any new development must meet. The result is a raft of paradoxical, self-contradictory views. They protest the use of genetically modified foods while insisting organic food can feed billions of people without contributing to climate change in the process. They revile the use of fossil fuels of any kind, but advocate government subsidizing of green technologies that only half work, short-circuiting the process of developing new technology before it can even start.

Thus far we have looked at progressive ideas and oddities in isolation. But by now it should be clear: the values behind today's progressive positions are, at their heart, unworkable as policy. Imagine an ancient civilization with a booming population facing the problem of dwindling herds of wild game and an insufficient supply of berries to forage. Then imagine that policy makers' hands were tied by today's progressive bugbears: agriculture would be a non-starter—too disruptive to the environment. Without any way for conflicting prehistoric stakeholders to increase their food supplies, these ancient progressives would be forced to fall back on rationing and mitigation—and on population reduction, in the worst-case scenario. Fortunately for us, ancient scientists did not embrace progressive mitigation and rationing. Instead, they invented agriculture and domesticated livestock.

Look forward a few centuries. Would the introduction of pasteurization have passed muster according to the precautionary principle? Not if you were a raw milk proponent. Eradicating bovine tuberculosis and typhoid fever in people was hard to object to, but the precautionary principle would have contended that pasteurization hurts small milk producers and dries up the milk supply, thereby hurting everyone. Dangerous milk was better than less milk, they would have asserted. In a 1940s speech before Parliament defending pasteurization, Lord Rothschild summed up the arguments of the opponents of pasteurization in words that would sound just as likely coming from today's Whole Foods devotees: "What nature produces is good enough for me, so better not tamper with it; it might be dangerous; I like my milk raw."[1] Raw milk proponents were wrong then and remain so today. Centers for Disease Control statistics show that raw milk is responsible for nearly three times more hospitalizations than any other foodborne disease outbreak in America.[2]

Besides coming up with a method for killing milkborne diseases and vastly ameliorating infant mortality, Louis Pasteur's work also contributed to the development of the first vaccines and debunked the then-prevalent idea that disease sprang from spontaneous generation—the idea that living things can arise from nonliving things, which we now know is a violation of cell theory. Perhaps it is a coincidence—perhaps not—but consider: who initially propagated anti-vaccine ideas? Progressive activists, who started arguing a decade ago that vaccines caused autism and then subsequently deployed anti-science fallacies to shrug off data to the contrary.

The impulse to impose an unworkably utopian vision on society has been part of the progressive movement from its earliest days. Today's progressives would prefer to forget some of its earlier manifestations—namely, one particular old pet cause, long buried and hidden from polite society: eugenics. Founded by Charles Darwin's nephew, Francis Galton, the movement proposed "to check the birth rate of the unfit instead of allowing them to come into being. . . . The second object is the improvement of the race by furthering the productivity of the fit by early marriages and the healthful rearing of children."[3] In other words, this was a movement whose goal was to optimize human natural selection by sterilizing those less fit or less educated.

Distasteful as these ideas have become today, many leaders of the early progressive movement were avowed advocates of eugenics. Margaret Sanger, an early advocate and popularizer of birth control, promoted eugenic ideas.[4] Nobel laureate in literature and Academy Award winner George Bernard Shaw, a legend of the English left, put it this way: "The only fundamental and possible Socialism" was "the socialisation of the selective breeding of Man."[5] We should all count ourselves fortunate that in today's world eugenics has lost its allure. (Though as we shall see, there is still a trace of eugenical thinking on the fringes of progressive debate—at least when it comes to disliking those with whom they disagree.)

Between the various self-defeating causes that progressives have taken up (and sometimes discarded) over the years, what ultimately becomes clear is that today's progressives combine elements of the worst of two different worldviews. They are happy to latch on to any suspect social science study that reaffirms their core beliefs, yet they remain highly resistant to advancements that fall outside their political and cultural comfort zone. This mentality is the antithesis of transformative science. A progressive assault on jarring, uncomfortable advancements is a war on the nature of science itself.

As we approach the end of our adventure into the world of anti-science progressivism, it is worth returning to the principles that started our journey: the values at the core of progressive thought. To do that, we have to distinguish how people's values don't easily fall into left and right halves of a graph, or even into quadrants.

We're Not Left and Right; for Each Issue We Are a Triangle

Our very first act in this book was to lay out the four categories that inhabit the grid of today's political model—liberals, libertarians, progressives, and conservatives. This is a useful model—but it goes only so far. Obviously, in everyday life real people contain multitudes of sometimes contradictory ideas. Liberal is a limited term, just like libertarian is. People are far more nuanced than left and right. As we noted, a New York liberal and a San Francisco progressive don't have much in common beyond voter registration. There is an even wider divide within the broader community of progressives, just as there is within the broad rubric of conservatives.

Life and policy are about priorities. Whether it is more important to spend tax dollars on missiles or on education. Whether it's more important to preserve a barren swath of tundra or power a nation. The same is true of politics. What matters most in terms of defining these philosophies is their prioritization of values.

We began this book by taking cues from the quadrant system of values. But in order to find our way to the essential clash between progressivism and science, we need to revise this system. We have found that a triangle may be an even better way to think about political issues. Each issue comes down to striking the right balance among three core values.

The very etymology of "liberal" and "libertarian" stems from the Latin word *liber*, or "free." Liberals and libertarians simply disagree on the degree of liberty individuals should have. In our classification system, they don't need to be separated—both are similarly preoccupied with freedom.

A core tenet of progressive ideology is social justice and equality—leveling the playing field. "Fairness" is therefore another suitable label.

Conservatives often claim that self-determination and personal initiative are the paramount virtues. This is a little more complicated than the universal concepts of freedom and fairness, so we'll simplify by saying conservatives want people to be able to excel if they choose. Thus, "excellence" is a positive word, and it also happens to be the most important component of good science. The best researchers and the best science should win. Nobody objects to excellence (except maybe losers), and everybody loves winning—particularly Charlie Sheen.[6]

That leads us to a triangle-shaped playing field. Where you come out on an issue—whether you prioritize freedom first or

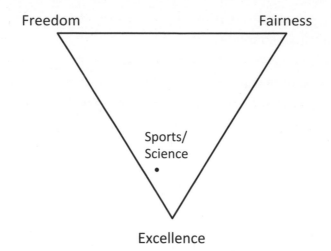

whether fairness or excellence is most important to you—dictates where you end up on the triangle.

If you occupied the exact middle of our triangle, it would mean you cared equally about freedom, fairness, and excellence on any particular issue. You would be the perfect moderate—heeding all sides to perfection. The reality, however, is that we all make trade-offs that dictate which point on the triangle we are closest to.

Take the sport of professional football. If you were creating rules for football teams in an alternate universe, would you make freedom the prevailing value and avoid regulating the sport at all? If so, you might end up with a serious steroid problem, as players would do whatever they pleased to win. Or you might end up with considerable injuries, as team owners would work their players beyond the point of exhaustion, answering to no one.

Alternately, you could rule over your imaginary sporting universe by dictating that everything should be fair before anything else. All players would have the same size and skill, and the coaching playbooks would be the same. The result would be a pretty boring sport.

Yet that isn't to say that excellence alone is desirable. Fairness in opportunity is a must, as is freedom. Though one value takes precedence for different people regarding different issues, the balancing act is crucial.

We believe that science is no different from sports when it comes to the triangle of values. Freedom is important because scientists need to be free to investigate the laws of nature and fundamental issues—even uncomfortable ones that put our deeply held assumptions and beliefs to the test. Fairness in opportunity is essential, but it cannot be the sole quality. While we want there to be no artificial obstacle to science excellence—like prejudice for height or hair color or race determining who gets jobs—artificial fairness would mean all scientists would get the same grants and there would be no competition. Without competition, there would be no revolutionary, transformative science. Thus, we believe the best science prioritizes excellence, followed by freedom, and then finally fairness—exactly in that order—and the weighting toward that node of the triangle shows it.

The foundations of science begin to shake, however, when science is skewed in directions that have nothing to do with excellence. We might even begin to make the case that progressives are engaged in an undeclared war on excellence itself.

Fashionable Nonsense
Is a Fabric of Many Threads

Hijacking the culture of the people who make up science is tantamount to hijacking science itself. Science, as a strategic resource, needs to be a meritocracy. That means the best people will shine, regardless of color, creed, or religion. If progressive special interests were allowed to demand representation equal to its desires (or more!), the nature of discovery itself is corrupted.

As we have argued, in science, excellence is paramount. People need to be able to take the initiative to do groundbreaking work. Most Americans recognize that part of the reason America leads the world in science is that we have a culture of freedom and a meritocracy that rewards performance. Fairness is important only in the sense that people should not be unfairly blocked out of a field. Thus, the most vital components of quality science are individual initiative—a conservative trait—and the freedom to go after it, a liberal/libertarian one. Religious dictatorships, for example, produce very little great science because scientists lack the freedom to create it.

In many respects, science is no different from athletics. You don't get to be called an elite athlete unless you outperform your competitors. People who want to watch great football place themselves near the excellence point on our triangle. They want to see the best players; they do not want arbitrary limits and regulations imposed by fiat. Nobody wants to see, for instance, rules mandating that each ethnic group receive representation in the National Football League (NFL) equal to its overall population numbers. Not a single NFL fan cares about demographics; everyone cares about performance. So it goes with the broad science community. However, many progressives are indeed more interested in promoting their own special interests—sometimes by fiat—over the far more important need of excellence in science.

For instance, many progressive commentators have protested the lack of representation of women in the sciences. They've mustered all manner of studies to support this contention. You see it in sociological studies that are sent out for media consumption every month; recent ones imply women are so emotionally fragile that if equal numbers of men and women aren't in a classroom, they are threatened by inequality and suffer "gender fatigue."[7]

(This is a phenomenon described by social psychologists wherein the opposite sex—usually men—tires of being told how sexist it is and therefore indirectly promotes sexism by doing less to combat it).[8] Some contend women are so worried about a supposed reputation for not being good in science and math that they score more poorly in science and math due to that pressure, a vicious psychological circle.[9] We find these arguments contradictory at best and patronizing at worst. Feminism correctly taught all of us that women are strong, independent beings perfectly capable of succeeding on their own. Now, progressives are telling us the opposite: women shrivel when confronted by big bad men. Instead of encouraging women to be strong in society, the modern progressive knee-jerk response is to blame everyone else.

To look at the landscape of women in science objectively, one must first deal with two observations. First, women, on average, are paid less than men. Second, in some jobs there are fewer women than men. There's no getting around those issues, no matter how the data is presented.

But what is left out of progressive complaints about female representation is that the disparity changes dramatically when you include the social sciences. In the social sciences and in fields like education, 71 percent of the workforce is made up of women.[10] Yet you'll rarely hear anyone make the claim that men face a hostile environment in psychology classes or do poorly because of stereotype threat or gender fatigue. Meanwhile, no one on the left bats an eyelash at the staggering 6:1 ratio of Democrats to Republicans in overall university faculties[11] or an even more shocking 10:1 ratio in biology and 44:1 ratio in sociology. Also, the few remaining endangered Republicans tend to be older, meaning they aren't being hired for new jobs that open up. Apologists invoke self-selection bias to contend that this under-representation simply means that Republicans have chosen to do other jobs . . . or, worse, that they are too stupid to be academics. You also don't hear anyone griping that constant sniping and jokes about Republicans create a hostile work environment for them.

The *New York Times* detailed an example of rationalizing away the lack of conservatives in academia. At a social psychology conference, Dr. Jonathan Haidt, who self-identifies as a liberal-turned-centrist, conducted a survey of the audience:

> He polled his audience at the San Antonio Convention Center, starting by asking how many considered themselves politically liberal. A sea of hands appeared, and Dr. Haidt estimated that liberals made up 80 percent of the 1,000 psychologists in the ballroom. When he asked for centrists and libertarians, he spotted fewer than three dozen hands. And then, when he asked for conservatives, he counted a grand total of three.
>
> "This is a statistically impossible lack of diversity," Dr. Haidt concluded, noting polls showing that 40 percent of Americans are conservative and 20 percent are liberal. In his speech and in an interview, Dr. Haidt argued that social psychologists are a "tribal-moral community" united by "sacred values" that hinder research and damage their credibility—and blind them to the hostile climate they've created for non-liberals.
>
> "Anywhere in the world that social psychologists see women or minorities underrepresented by a factor of two or three, our minds jump to discrimination as the explanation," said Dr. Haidt, who called himself a longtime liberal turned centrist. "But when we find out that conservatives are underrepresented among us by a factor of more than 100, suddenly everyone finds it quite easy to generate alternate explanations."[12]

The same pitfall faces science as progressives continue to engage in a cultural tug-of-war, lobbying for various constituencies and mocking conservatives without heeding the effects on science itself. Artificially handicapping the research community in the name of progressive good works won't lead to better science. It will lead to a eugenics for science, where all serious thinkers will find themselves neutered.

Why It Matters

Obviously, a liberal environment in the classical freedom sense is a good thing for science, but today, for the first time in world history, academia is instead almost entirely controlled by progressives. They are pushing artificial fairness above true excellence, thus damaging the scientific enterprise.

William F. Buckley originally noted his concern about changing currents in academia in 1951. His book *God and Man at Yale* was written as a reaction to what he saw as forced ideology at his alma mater. Professors had set out to kill individualism, he wrote. He thought that what he was witnessing was a revolution in which conservatives were taking up the mantle of freedom abandoned by liberals. Little did he know that he was dealing with the emergence of a different breed of liberal—what we have now classified as progressives.

In the decades that followed, as anti-science and cultural agendas began to encroach on academia, people interested in liberal freedom left in greater numbers and were replaced by like-minded progressives.[13] Academic science gradually became more representative in race and gender, but far less in politics. Were Buckley alive today, he would be horrified to learn how Republicans have been ostracized. The liberals of his day would never have gone to such extremes to squelch diversity, but their progressive descendants have. Making the problem even worse is the fact that these progressives have captured the policy-making, funding-deciding apparatus—which means they are in charge of deciding priorities for the next generation of academics.

Now, extreme progressive authorities have declared that their ideological opponents are actually inferior, even trying to make the claim that there is an evolutionary basis for conservatism that makes its adherents less trustworthy as scientists than progressives are. Liberals have brains that make them the kind of people who like to hang out in places like universities and try out new ideas, while conservative brains are too close-minded and defensive. Biologically, liberals just belong in universities. In truth, the bulk of the scientific community vehemently disagrees with such claims. Famed evolutionary biologist and professor Jerry Coyne, author of the blog "Why Evolution Is True," takes progressive author Chris Mooney to task for relying upon suspect correlations, like studies that measure skin conductance differences between people when they look at pictures of the Clintons, supposedly proving that political opinions are due to having a right-wing brain.

In Coyne's words, "Mooney concludes, then, that liberals are a bunch of soft-nosed tree-huggers and bunny lovers, while conservatives are alert and wary, easy to perceive threat. Where does the evolution come in? Because Mooney suggests that those differences, to the extent that they're genetic, arose by natural selection. Not only that, but 'liberal' genes are less adaptive than 'conservative ones!'"[14]

No rational person would contend that behavior is not impacted by biology. Of course it is. One day, we may even be able to better explain how biological factors combine with environmental ones to produce behavioral outcomes. But we are nowhere near that point today. And furthermore, using science to dissect the left-right political divide along specifically American political issues is downright silly. Conservatives, liberals, libertarians, and progressives believe different things in different countries. Additionally, political beliefs aren't static; Democrats from the 1950s don't even closely resemble today's Democrats.

The same is true of Republicans. Trying to use science to explain these political differences—especially when the goal is to smear your adversaries—is nothing short of scientific malpractice. Besides, this isn't the first time progressives have tried to map biological data to their cultural topology; it is just a modern-day variation on the same eugenical theme that the early progressives were so fond of.

Thus, the idea that conservatives believe what they believe because their brains are malformed is political nonsense at best and dangerous at worst. To scientifically literate people, it is obvious that genes don't make us vote. Likewise, it is obvious that political beliefs and social issues should have no bearing on scientific decision making. That's why it is crucial that, with scientific excellence as our underlying goal, we move beyond these petty, anti-scientific debates and address the issues that really matter, as we will in the remainder of the book.

Chapter Fifteen

TWELVE ISSUES FOR 2012 AND BEYOND

The Science That Really Matters

To JUDGE FROM THE SAME OLD tired debates that pop up every election cycle, there are only three controversies where science clashes with politics in this country. On the news, in papers, and in blogs, we get the impression that evolution, human embryonic stem cells, and global warming are the extent of the issues in the science culture war. Each politician—particularly if he is conservative—is judged as being "pro-science" or "anti-science" solely based on his personal stance on those three issues alone. For all the many vitally important topics that science and politics face, the media appear to be interested only in the science issues that make conservatives look especially bad while ignoring all the issues that make progressives look bad.

To take the most flagrant example, evolution, we'll put it bluntly: it's not a national issue, even if that hurts the feelings of a lot of our friends. It doesn't matter if Sarah Palin believes in it. Honestly, it doesn't really matter if the president of the United States believes in it either. Why? Because detailed decisions about school curricula are not made at the federal level; instead, they are made at the state and local levels. If a local school board is stacked full of creationists, then that should be a concern. But even then, the silly protracted "debate" over evolution is little more than a minor nuisance when compared to much more serious issues such as energy production, biomedical research, and public health. Remember, nobody has died from denying evolution, yet millions have died from diseases that vaccines now prevent and that crop up again when progressives do not take a stand against their own anti-science contingent.

Besides, who among us is perfectly rational all of the time? Jimmy Carter and Dennis Kucinich claimed to have seen UFOs. Nancy Reagan believed in astrology. Hillary Clinton had imaginary conversations with Eleanor Roosevelt. Barack Obama believed spending billions on clean energy subsidies would cause the earth to heal. Most, if not all, people believe and do weird things. Not all of them matter. If a politician denies evolution, it makes him look unintelligent, but its overall impact on public policy is minimal. Any self-motivated student can learn from a biology book without fearing what his local representative thinks about Charles Darwin.

So what issues *do* matter? The following is our list of twelve issues that will shape the future of science and technology in the United States for decades to come.

1. Funding science in an age of austerity. You might have noticed that the global economy hasn't been doing well recently. Countries throughout the developed world are struggling with large national debts. There is tremendous pressure on the U.S. government to reduce spending and enforce fiscal discipline. Difficult financial decisions will have to be made. Because of that, scientists should realize that federal science funding might not always be maintained at current levels. It might drop across the board, as it did during the Clinton years. They should hope (and lobby) for the best, while preparing for the worst. That means prioritizing research and adjusting budgets accordingly. If they don't, Senator Tom Coburn (R-OK) will gladly do it for them.

Coburn is both a medical doctor and a fiscal hawk. He understands the necessity of basic research, but he also is frustrated by wasteful government spending. Thus, he regularly publishes a list of what he perceives to be questionable federal investments, and some of that includes science. For example, he criticized the National Science Foundation for funding what became known as "shrimp-on-a-treadmill," a study that examined how shrimp react to changes in water quality.[1] Regardless of whether his attacking this particular study was fair, the general point Coburn made is that some federal money for scientific research is being wasted—and he was right. While the running shrimp had some value, taxpayer funding to see if Farmville helps build personal relationships does not, yet the NSF funded it for $300,000. There are many such examples. Few scientists would disagree that some waste will happen. Government programs are not known for efficiency, and funding for scientific research is not an exception to the rule.

Unfortunately, federal funding for science is a zero-sum game. One project that is awarded money means that another (potentially better) project did not get funded. As a result, scientists *must* be willing to admit that not all science is created equal. Some projects are more worthy of taxpayer funding than others. Instead of circling the wagons and claiming that all science is under attack when politicians like Coburn reveal waste, scientists should seek better ways of policing themselves and promoting the best research—and then apply for the extra money that would be available if sociologists stopped getting funded to play EverQuest 2. (We hear it's an awesome game, though.) It would also help if scientists more frequently emerged from their academic cocoons and explained to the public the importance of their research. Otherwise, if they don't, they might be subjected to public scrutiny they may not find enjoyable.

What sort of science funding should be cut? Certainly the funding that is not science but that is, instead, just subsidies for technology. The Solyndra experience has demonstrated that technology subsidies are ripe targets for elimination. Funding basic research in solar power is a perfectly legitimate use of taxpayer money, but subsidizing solar companies (especially ones with dubious prospects in the marketplace) is not. Using the Solyndra scandal as an example, the *Economist* avers that solar subsidies are "massively wasteful and squalidly political."[2] The federal government should also immediately close the embarrassing National Center for Complementary and Alternative Medicine, a pathetic monument to anti-science quackery created by progressive Senator Tom Harkin. Beyond that, scientists need to start thinking like the politicians who want to shut them down and recognize that their work has to demonstrate its value—or face cuts in the age of austerity.

2. Developing a smarter energy policy. Current U.S. energy policy makes absolutely no sense. Writing in the *Dallas Morning News*, Dr. Michael Webber puts it best:

> The energy debate that has raged for decades still breaks down into two ideological camps: those who believe in low production and low consumption [Democrats] and those who believe in high production and high consumption [Republicans]. Consequently, America has the worst of both—high consumption combined with low production.[3]

We agree with this assessment. So, how should the United States proceed?

First, progressives need to get over their knee-jerk opposition to fossil fuels. Culture war rhetoric aside, everyone wants a less-emissions-heavy, more renewable energy supply. But we simply cannot stop consuming fossil fuels overnight. Projects taking advantage of the resources on this continent (like the Keystone XL pipeline) should proceed. Progressives must also stop their anti-science crusade against natural gas. At one time, environmentalists embraced natural gas as the "bridge to the future," a transition to a clean energy economy. Not anymore. Instead, they downplay the fact that the United States is awash in natural gas and exaggerate the dangers of fracking—a technology that has been around since the 1940s.[4] Why this sudden turnaround? According to Jon Entine,

> Overflowing supplies destroy Big Green's argument that fossil fuels will get more and more costly till even wind and solar power are competitive. That undermines the argument for massive subsidies of alternatives that may never deliver competitive bang for the buck. No longer is natural gas a bridge to the alternative energy future. Much to the chagrin of energy activists, natural gas now is the future.[5]

Basically, progressives favored natural gas until it became so cheap and available that it threatened their visions of a green utopia. That's no way to achieve a sustainable energy policy.

At the other end of the spectrum, conservatives should drop their tiresome appeal to the almighty power of the "free market" as a catchall solution to every problem. There is no free market in energy. Fossil fuel companies are indirectly "subsidized" if they get special tax credits, just like clean energy companies are subsidized directly. Even though we strongly believe in the utility of the free market, it cannot solve everything. Carbon emissions are an externality (a problem, by definition, that the free market cannot solve) that must be dealt with, so perhaps a carbon tax could help decrease CO_2 while simultaneously making clean energy more competitive. Also, because nuclear power is expensive, it is not unreasonable for the federal government to play a larger role in assisting with construction costs and waste disposal. Conservatives often point to France as the prime example of a country that successfully embraced nuclear power—the dirty little secret here is that the nuclear power plants are largely owned by the French government.[6] That's one possible way forward, but conservatives are going to have to overcome their ideological resistance to government first.

While the United States transitions away from coal and oil toward natural gas and nuclear, the country should continue to invest in basic clean energy research. We are optimistic that a breakthrough in solar technology will happen, perhaps in our lifetime. (A free insider tip: keep an eye on solar research involving "quantum dots.")[7] There may also be a breakthrough in fusion power in our lifetime. Serendipitously, as technology marches forward, a sensible energy policy will automatically make for a smart climate policy. We really can kill two birds with one windmill.

3. Deciding the future of America in space. In 2011, NASA ended its thirty-year space shuttle program. Now the deep irony of America's victory over the Soviet Union in the space race is that, at least for a while, NASA astronauts must hitch a ride with the Russians aboard their *Soyuz* spacecraft. Thankfully, it will not always be that way. Private companies, such as Richard Branson's Virgin Galactic, soon will be leading the way into space. This will allow space tourism and perhaps even faster commercial airline flights.

But, really, that is just child's play. Americans want to go to Mars. Newt Gingrich wants to make the moon the fifty-first state.[8] Who will pay for these ambitions? Surely, private companies cannot afford to chase after wild dreams if there is no return on their investment. Should the federal government offer a prize to the first company that successfully reaches Mars, or should NASA be charged with going? Currently, NASA plans to go to an asteroid by 2025 and then to Mars by the 2030s.[9] Can we afford that, or is the era of Big Science over? There are no clear answers to these questions. Especially during times of tough economic decisions, it is imperative that Americans have a serious debate about the country's future in space.

4. Expanding human embryonic stem cell research. Yes, this is one of the three issues that actually *does* get discussed—but now we need to start talking about it as adults.

President Obama's executive order allowing federal funds to be used to create new human embryonic stem cell lines made his supporters feel good, but it had little practical effect. (See Chapter 2.) But that didn't stop progressives in the media from acting like Obama had sprinkled American science with holy water (secular holy water, of course). In reality, instead of opening the floodgates, Obama's new policy merely nudged the door of hESC research open a little farther than it had been under George W. Bush.

There is a policy, however, that could change hESC research overnight. Fertility clinics all over the nation routinely practice in vitro fertilization (IVF). Normally used for couples who have trouble conceiving, the procedure involves combining human eggs and sperm to produce zygotes (miniature embryos) in the laboratory. Several zygotes are produced, and a few are implanted into the woman's uterus. The remaining zygotes are often stored in a freezer, just in case the parents need them for the future.

This practice presents an enormous bioethical inconsistency. Many Americans oppose hESC research because it destroys embryos, yet the non-controversial and routinely practiced IVF technique essentially does the same thing. The unused embryos are often frozen indefinitely, never to be seen again. It is estimated that hundreds of thousands of such embryos sit frozen in time as a result of this technique.[10] What should be their fate? Obama's executive order allows them to be used for hESC research if the parents give informed consent. However, this regulation does little to address the hundreds of thousands of embryos that were frozen before Obama became president. What should happen to them? Some parents offer them to other infertile couples, some pay money to keep them frozen forever, and others choose to thaw and discard them.

If an Obama (or any other subsequent) administration actually wanted to address this issue, the latter two options should not be allowed. If after a few years frozen embryos are unused, they should become eligible for hESC research with or without permission from the donors. Literally overnight, thousands of zygotes would be removed from suspended animation and made eligible for scientific research. This is much more ethical than permanently freezing the embryos or throwing them in the trash.

Furthermore, the NIH should allow federal funding for somatic cell nuclear transfer. This procedure transfers a nucleus from a body cell into an egg donated by a woman. The technique was used to create Dolly the sheep, and in humans it would be used to study therapeutic cloning and regenerative medicine. Yet it is not eligible for NIH funding. This policy should be reversed immediately, sidestepping the tired culture war and focusing on what really makes a difference.

5. **Balancing technology with privacy concerns.** Privacy is something most people, particularly Americans, guard very jealously.

Many feel that governments and businesses should know as little information about us as possible. Yet as technology advances, the ability of others to collect private data about us becomes easier. For instance, websites already track your activity online in order to more effectively target advertisements just for you. If you read sports news online, it may not be a coincidence that the ads you see are trying to sell you gear from the local college team. And those ads about rejuvenating your hair and reenergizing your erections probably aren't a coincidence either.

Technoethics will become more relevant as the Internet generation gets older. Everything we do online is recorded, often indefinitely. Did you say or do something embarrassing online? Too bad. Somewhere, out in cyberspace, it is probably cached. And what should happen to your online persona (e.g., your Facebook profile) when you die?[11] Should it be memorialized? Deleted?

Advances in biotechnology are causing bioethical headaches. For instance, the ability to read somebody's mind soon may no longer be just a science fiction fantasy. Scientists have shown that it is possible to reconstruct mental images from a person's brain using functional magnetic resonance imaging.[12] But that will only be a concern far down the road. Right now, it would be possible to create a national DNA database that contained a DNA "fingerprint" from every person in the country. This would make crimes such as child kidnappings and murders easier to solve. If all citizens had a national biometric identification card, illegal immigration would be reduced. Yet many people are frightened by the idea of Big Brother having this much knowledge and power.

Technoethics and bioethics present very difficult challenges. Determining the proper balance between society's needs and an individual's right to privacy should be part of the national conversation now—because that balance is only going to become more of an issue as technology becomes an even more intimate part of our lives.

6. Securing sensitive information. The H5N1 influenza virus, un-like the seasonal flu, has a high case-fatality rate. (This refers to the percentage of people who die after becoming infected.) Some experts think it is as high as 60 percent, whereas others suggest 14–33 percent.[13] Regardless of the exact number, it is very high. The one saving grace of the virus is that it is not easily transmitted between humans. So in an effort to understand how a virus evolves contagiousness, scientists introduced mutations into the H5N1 virus and created a strain that was easily transmissible between ferrets. Because ferrets are used as a model for human infection, the research could have created a virus that is also highly contagious among people. Essentially, scientists engineered what could be a highly deadly, highly infectious influenza virus that had not existed before. This sparked an enormous controversy within the scientific community and the public at large.[14] Should this data be released? (Ultimately, the decision was to release the data.)

The benefits of the research are obvious. To fully understand how influenza pandemics occur, scientists must study the genetics of the virus. However, the dangers of the research are obvious as well.[15] What if the virus is accidentally (or purposefully) released? What if another country with a biological weapons program uses this information to engineer an attack? (The likelihood of a terrorist organization creating an enhanced influenza virus is quite minimal; the skill and expertise required to manipulate a virus in the laboratory are substantial.) Supporters of such research often respond by indicating that nature has already produced dangerous organisms all by itself—from smallpox to antibiotic-resistant bacteria. They contend that, eventually, nature would have generated this highly contagious H5N1 virus anyway. Regardless of where you stand on this issue, determining whether certain research should be allowed or if scientific knowledge should be censored will be hotly debated as science continues to advance into uncertain—and sometimes dangerous—territory.

7. Determining the appropriate amount of regulation. If progressive ideology could be summarized in a single word, it wouldn't be "progress"; it would be "regulation." Regulation (and its evil twin, litigation) is key to understanding progressive politics. Nothing is too insignificant to be watched over by Big Brother. That's why progressives are the ones who ban goldfish,[16] take away our plastic bags, and cause costume capes for children to carry the warning label "Cape does not allow the user to fly."[17] Progressives assume that corporations are bad and that most people are too stupid to take care of themselves, so the government must step in on their behalf. Of course, this is all done in the name of protecting the public. If anyone disagrees on scientific or economic grounds, progressives are quick to label him or her a shill for Big Industry.

Does regulation actually protect people? Some of it does; some of it is more for show. The *Economist* (a British publication if you're unfamiliar with it) believes that the United States is way too over-regulated.[18] Adding to this, Dr. Henry Miller, who formerly worked at the FDA, believes that the agency does not promote innovation.[19] Despite this, progressives insist we aren't regulated enough, and they focus only on the positive aspects of regulation while ignoring all the harm that it does. Increased regulation stifles creativity and chokes innovation. That is not just economic theory found only in conservative- or libertarian-leaning economic textbooks. On the contrary, it happens in real life: over the past few years, venture capital firms have pulled back their investment in biotechnology due to fears over increased FDA regulation.[20] Determining the appropriate balance of regulation—on the one hand to protect the public, but on the other hand to not discourage technological innovation—is a debate that is vital to the future of science in America.

8. Promoting sound public health policies. Americans take good public health for granted. Our life expectancy is high because we all have easy access to food, and infectious disease is merely an afterthought for most people. Today, our biggest concerns are obesity, cardiovascular disease, and cancer, all of which are associated with lifestyle choices. Indeed, life in the developed world has never been easier than it is today.

However, we must never become complacent. Hurricane Katrina reminded us that we are one disaster away from a total breakdown in law and order, including public health. Following the hurricane, diarrheal diseases occurred as sewage systems failed.[21] This should serve as a warning that infectious diseases never "go away." That is why we continue to vaccinate. Nasty bacteria, viruses, and parasites are always lurking around, just waiting for a crack to develop in our public health armor. Also, nature isn't our only potential enemy: terrorists could use chemical and biological weapons to wreak havoc on society, and a strong public health system would be our first line of defense.

Dr. Michael Osterholm believes that modern public health is founded upon three pillars: vaccination, chlorination, and pasteurization. To those, he has suggested adding a fourth: food irradiation.[22] The problem, of course, is that the word "irradiation" scares a lot of people, particularly progressives, so we would have to call it "zapping." To zap food, we would place it in something like an X-ray scanner and give it a brief dose of high-energy electromagnetic radiation. This is necessary because bacteria can hide inside food, so hygienic agricultural practice or scrubbing of vegetables properly is not always sufficient to eliminate them.[23] Perhaps if this technique were widespread, the deadly European *E. coli* outbreak in 2011 would never have happened. Also, consider that about 48 million Americans are sickened by foodborne illness each year, resulting in 128,000 hospitalizations and 3,000 deaths.[24] Food irradiation would not only save thousands of lives; it would also likely save billions of dollars in lost productivity and health care costs.

9. Balancing open access to scientific information with intellectual property rights. Currently, there is an ongoing revolt among scientists in regard to the publication of scientific information, and the community has aimed much of its anger at a publishing company called Elsevier. Elsevier isn't unique; it just happens to be really big and powerful. Scientists are upset because they claim large publishers charge too much for articles, force libraries to purchase bundles of journals (which often contain worthless ones nobody reads), and work to restrict the free exchange of information.[25] Further, because a lot of research is taxpayer funded, Elsevier and other publishers are being indirectly subsidized by the government and then charging scientists to read information that was provided to the publishers for free. (Scientists provide the data to the publisher and even do the editing work for them, often without being paid a nickel.) It's a nice business model for Elsevier—which cashes in with a 36 percent profit margin—but infuriating for everybody else.[26]

Macmillan Publishers, the conglomerate that owns *Nature*, benefits from the same indirect subsidies as Elsevier, but what got scientists really angry at Elsevier was its support (later withdrawn) for a bipartisan piece of legislation called the Research Works Act. Currently, most life sciences research that receives taxpayer dollars must be made available free of charge a year after publication. This allows a journal to profit from exclusive rights to scientific information for twelve months, but after that the information must be free for everybody. That is probably an acceptable compromise. However, the Research Works Act would have done away with this open-access policy.[27] Scientists responded by boycotting Elsevier and urging the entire community to use open-access journals.

Similar to this is the touchy issue of gene patenting. A company called Myriad Genetics has filed several patents related to testing for mutations in what is popularly known as the "breast cancer gene," BRCA.[28] The effect of the patents is that Myriad controls the market for BRCA screens. Supporters of gene patenting claim that protection of intellectual property is vital for companies to develop new technologies. Without them, anyone could steal their hard work, and thus biotech companies would lose money and go out of business. Opponents claim that gene patents hinder basic scientific research and prevent others from attempting to develop alternative technologies. Determining the proper balance between these two positions is extremely difficult. There are merits to both sides of the argument, and currently gene patenting is being adjudicated in the courts—in March 2012 the Supreme Court overturned a lower appeals court ruling, sending the case back down to be fought again.[29] Indeed, this issue will be a hot topic for bioethicists and patent lawyers for years to come, and whatever policy is finally agreed upon will have an enormous impact on the future of science in America.

10. **Managing resources efficiently.** The population bomb scaremongering of Paul Ehrlich has been defused. Global demographic studies have consistently shown that as societies become more educated and productive, birth rates tend to fall on their own.[30] The solution to overpopulation—which is a problem only in the minds of progressives—is to encourage technological advancement. Inhumane initiatives, such as China's one-child policy, are counterproductive, biased against baby girls, and completely unnecessary.

However, this does not imply that there is no room for improvement. The world needs to be better at resource management. Fred Pearce believes that overconsumption is a far bigger environmental threat than overpopulation.[31] There are two general approaches to solving this problem: reducing consumption and increasing efficiency. The problem with the former is that people agree with it in theory, as long as others decrease their consumption. (Similarly, most people support taxing "rich people," as long as they don't see their own taxes increase.) The latter option, increasing efficiency, we described in Chapter 4. The world can produce more goods using fewer raw materials. This is much more appealing, and in many ways the world is already trending in this direction.[32]

Ironically, environmentalists who oppose nuclear power and favor organic food are actually supporting less efficient methods of production—which is ultimately worse for the planet.[33] If we want the world to be more prosperous, we must embrace every new technology that increases efficiency. That includes nuclear power and genetically modified crops. Progressive environmentalists must put aside their technophobic tendencies and embrace change—which is supposedly what "progressive" stands for anyway.

11. **Addressing global poverty.** We all know how annoying it is when the electricity goes out for a few hours. We can't use the computer, we can't read at night, we can't see in the bathroom, and, most disappointingly, we can't keep up with the Kardashians. Now imagine *never* having electricity. Without it, a country cannot advance technologically or economically. Food spoils more quickly without the electricity required for refrigeration. And it is difficult to sustain good public health since hospitals and water treatment facilities require electricity. Though we take

electricity for granted, the reality is that 1.3 billion people on the planet lack access to it.[34] Addressing this problem will require continued international investment, in addition to the flexibility of using any energy source available, including fossil fuels. Even then, oil may not save developing countries as its price climbs higher due to demand and geopolitical instability. As *Time* wrote,

> Environmentalists sometimes welcome higher energy prices as a spur to conservation and efficiency, but that's true mostly in rich countries; in developing, energy-starved ones, high prices can be economically crippling.[35]

Energy poverty isn't the only problem facing the developing world. Infectious disease still ravages many parts of the planet. Included among the leading causes of death in low-income countries are diarrhea, AIDS, tuberculosis, neonatal infections, and malaria.[36] To help combat this, President George W. Bush, whom progressives referred to as an anti-science president, tripled funding to Africa to combat AIDS, malaria, and tuberculosis.[37] And Bill Gates, a member of the allegedly greedy "1 percent," is using his vast wealth to fight "neglected" parasitic infections that most people in the developed world have never heard of.[38] The United States, fueled by great wealth and the generosity inherent in the American spirit, should continue the fight to bring the benefits of science to those who need it most.

12. **Restoring science to its rightful place.** In his 2009 inaugural address, President Barack Obama promised to "restore science to its rightful place." Instead, as we have detailed extensively throughout this book, he simply replaced conservative anti-science policies with progressive ones.

How do we *actually* restore science to its rightful place? A good starting point would be to fix science journalism, which we believe is fundamentally broken. The Fourth Estate once held *everyone* accountable, but that has been lost. Also, most newspapers and media outlets barely cover science. When they do, they tend to sensationalize and hyperbolize. They report stories without giving proper context, and journalists too often act like science cheerleaders instead of asking tough questions. Politically oriented bloggers and science writers—many of whom are progressives—have used their platforms to attack conservatives instead of discussing science. This has to stop. We wrote this book, not to defend conservatism, but to remind those who hurl unfair attacks against conservatives that the same label of "anti-science" easily applies to them as well.

The solution, then, is to quit politicizing science. That will be difficult to do, as some scientific topics have national policy implications. Scientists must do their best to make sure their data is reported accurately. If they also choose to become policy advocates, they must identify themselves clearly and be sure not to sensationalize their findings to make a point. In other words, they must not act like climate scientist Dr. Stephen Schneider, who once said,

> So we have to offer up scary scenarios, make simplified, dramatic statements, and make little mention of any doubts we might have. . . . Each of us has to decide what the right balance is between being effective and being honest.[39]

No, Dr. Schneider, that is not what each of us must decide. We must decide how to effectively *and* honestly communicate the lessons science teaches us.

The best way to do that is through transparency. A good example is the Berkeley Earth Science Temperature (BEST) study under the direction of physics professor Dr. Richard Muller.[40] To help settle the question of whether or not global warming was real, he very publicly analyzed 1.6 billion data points. In the end, he reaffirmed what climate scientists had been saying for years. Such diligence and transparency are good for the scientific enterprise and help reassure the public at large. Indeed, if more scientists, journalists, and bloggers were like him, that would go a long way toward restoring science to its rightful place.

CONCLUSION

Closing Arguments:
An Appeal to Scientific Reasoning

BOTH OF US HAVE BEEN INVOLVED with the science and technology fields for a big chunk of our lives. Combined, we have about thirty years of experience in the halls of academia and corporate boardrooms, with PhDs and with undergraduates. Throughout our various careers, we have met and worked with scientists who were liberals, progressives, atheists, Christians, Jews, Muslims, conservatives, libertarians, and even one self-identified Buddhist-Christian, whatever that is. We found across the board that people in science are generally willing to detach their cultural, political, and religious beliefs from their research. Worldview is mostly unimportant when it comes to the global scientific endeavor. People in science usually avoid the shrill activists in their ranks.

Accusations of a Republican war on science, of conservative resistance to and defiance of fact, do not contribute in any positive way to the scientific enterprise. These accusations are founded on a handful of scientific topics on which slightly more conservatives hold slightly more uninformed beliefs than those on the other side. It blatantly ignores all of the uninformed beliefs that those who aren't conservative hang onto.

Progressives routinely toss the anti-science bomb at conservatives. Their own anti-science beliefs get off scot-free—even when they conflict with those of other progressives. For example, global warming, anti-GMO, and anti-meat activist camps all spin their wheels, lobbying for different, even contradictory laws while not trusting science to solve these core issues. Meanwhile, the anti-science bomb has become so commonplace that the mainstream media don't even bother to check the accuracy of the people doing the dropping or the munitions it contains. Smearing conservatives on scientific issues has become a run-of-the-mill talking point—red meat for left-leaning politicians and pundits to serve up to their base and grist for the mill on op-ed pages and in the blogosphere.

Until now.

What makes somebody pro-science? A pro-science person trusts the scientific process and is not afraid of the truths science reveals. That doesn't mean that we should trust science blindly—we need to be wary of dubious health studies and research that makes shocking, but poorly supported, claims. We should always scrutinize studies for potential conflicts of interest, such as those pushed by pharmaceutical companies in their marketing campaigns. But, overall, the scientific method works—whether the research is conducted in universities or in big corporations. We need to trust in science when it reveals uncomfortable facts, even when they shake our political and religious beliefs to the core. When confronted with intellectual dissonance, a truly pro-science person must weigh the evidence and decide in favor of the side that is more persuasive. But beware: persuasiveness should be rooted in data, not emotion.

What, then, makes a person anti-science? First and foremost, a person who uses science as a facade for a philosophical or political agenda is anti-science, regardless of what they claim. Such a person is easy to identify: they usually have no training in science, they cherry-pick data to fit preconceived beliefs, and then they blog about it. An anti-science person prioritizes scoring political points over scoring scientific points. Also, a person who believes that dishonesty is rampant in the scientific process is by definition anti-science—by this we mean the people who dismiss academics as beholden to funding or who insist corporate scientists are routinely faking results for a paycheck. Some individual scientists are dishonest, but the scientific method is designed precisely to weed out inaccurate information. And in reality, we're *all* tainted by a need for money. (Have you ever seen the TV show *Survivor*?) In the end, however, the scientific method prevails.

If we're really honest with ourselves, the truth is that all of us can be anti-science at times. The reason is because we are not cold, calculating robots, nor are we über-rational ants in a heavily structured society; we are sentimental beings who are often too easily persuaded by emotional arguments. Everybody knows that airplanes are quite safe, but that doesn't prevent us from feeling anxiety when there is turbulence. Everyone knows that tobacco is unhealthy, but that doesn't prevent smokers from lighting up. All of us are subject to beliefs or behaviors that could be labeled anti-science. The key is to recognize them and then change our beliefs and behavior accordingly. This will make for a more scientifically literate and technologically savvy society as we venture through the twenty-first century. Don't let progressivism—or any other political ideology—prevent that.

In closing, we wrote this book to defend science and scientists from an unchallenged, pseudo-scientific, progressive onslaught against reason and data. We wrote this book to show how some of the basic assumptions among the progressive community are oftentimes just feel-good fallacies dressed up in the language of science. We wrote this book to expose the many anti-science beliefs that progressives hold, while they simultaneously claim to be the only true champions of science. We wrote this book to demonstrate that, upon further scrutiny, progressives' claims of being pro-science evaporate when science conflicts with a progressive agenda. We also wrote this book to reinvigorate the community of scientific journalists and writers to once again ask awkward questions and to stop being cheerleaders for science. Be trusted guides, instead. And perhaps most importantly, we wrote this book to call a halt to the scientization of politics and the politicization of science.

We would like to leave you with one final quote. When in a committee hearing on national defense, Representative Hank Johnson (D-GA) expressed some geologic concern about placing too many troops on the island of Guam:

> My fear is that the whole island will become so overly populated that it will tip over and capsize.[1]

Maybe in one of those alternate universes that astrophysicists tell us about, islands tip over, Frankenfish lurk in the oceans, and cell phones cause cancer. But not in this one.

For our part, if even one progressive has read this book and stopped worrying that the earth will deflate due to natural gas extraction, it's all been worthwhile.

ACKNOWLEDGMENTS

As rookie book authors, we really had no idea how much work it involves, not just for us but for an entire team of people who jumped in with enthusiasm to make us look good. As you would expect, there are far too many people to mention by name, but a few individuals deserve special attention:

Armed with a concept and a couple of sample chapters, our first stroke of good luck was finding the right agent. Craig Wiley was in love with this project from the moment he read it and tackled finding the right publisher. And he did, in PublicAffairs.

Our editor Brandon Proia is, in a word, superb. Despite his experience and our lack thereof, he treated us like we were a dream team of Stephen Hawking and George R.R. Martin. Then he took what was, in hindsight, rather raw, and skillfully weaved together a narrative from the collection of essays (rants?) that we delivered to him. He answered every question we had for five months (and we had a lot) but most importantly, he's a nice guy and fun to work with. He does, however, owe us lunch.

Lisa, Melissa, Emily, and everyone else we worked with have all been terrific. The entire PublicAffairs team has our sincerest thanks and appreciation.

Additionally, Alex would like to acknowledge the following friends and colleagues: Jeremy Lott, who was instrumental in helping to launch my writing career; David Boze, who gives great feedback and always promotes RealClearScience on his radio program; Rich Darveau, my former research mentor at the University of Washington, who has given me constant encouragement since my days as a graduate student; Jon Bean, who serves as a role model in standing up for one's beliefs in the face of overwhelming adversity; Deborah Illman and David Gordon, both of whom helped change the course of my career with their excellent advice and guidance in regard to science writing; Todd Myers, the eco-mythbuster, who generously provided an encyclopedic knowledge of all things environmental at a moment's notice; and my dear family and friends, all of whom provided encouragement, feedback, and when necessary, criticism. And of course, Hank, whose knowledge and sense of humor made the collaboration enjoyable. I couldn't have written this book by myself.

Hank would like to thank the many book authors, journalists, scientists, and contributors at Science 2.0. They're liberals, every one (but not progressives!), and when they heard about this book, they invariably said it needed to be written. Six years ago, with nothing but enthusiasm and a concept, I approached world-famous people about helping me reshape how science could be done, and they did it for free. In the process, they taught me how to write. Thanks mom, you remain the smartest person I know; Pete, Tom, and Vic, who have made the continued success of Science 2.0 possible; the Rebel Alliance, who helped me blow off steam taking down evil empires when work was done; my family and friends, who have always been convinced I am a "Goodfella" because they would visit my offices for fifteen years and no one ever seemed to be working, but everyone made money—see, here is proof I do stuff; and, of course, Alex. His enthusiasm made this a lot more fun than doing it alone.

NOTES

Introduction

1. Architect of the Capitol, "Going Green and Saving Energy," April 2010, http://www.aoc.gov/aoc/press-room/loader.cfm?csModule=security/getfile&pageid=114029.

2. Jonathan Karl, "Stick a Fork in It: GOP Defunds 'Green the Capitol' Initiative," ABC News, March 4, 2011, http://abcnews.go.com/Politics/gop-defunds-nancy-pelosis-green-capitol-environment-initiative/story?id=13022195#.T1G5oVHR3dl.

3. Architect of the Capitol, "Performance and Accountability Report," 2010, http://aoc.imirus.com/Mpowered/book/vaoc10/i1/p1.

4. Committee on House Administration, "Composting Program Suspended," press release, January 24, 2011, http://cha.house.gov/press-release/composting-program-suspended.

5. See http://twitter.com/#!/NancyPelosi/statuses/42376537084928000.

6. Felicia Sonmez, "House to Test Reusable Dishware," *Washington Post*, March 2, 2011.

7. Congressional Budget Office, *Federal Support for Research and Development* (Washington, DC: Congress of the United States, 2007), 11.

Chapter 1

1. Howard Bloom, *The Genius of the Beast: A Radical Re-Vision of Capitalism* (New York, NY: Prometheus Books, 2010).

2. U.S. Department of State, Smoot-Hawley Tariff http://future.state.gov/when/timeline/1921_timeline/smoot_tariff.html.

3. Stefan Kühl, *The Nazi Connection: Eugenics, American Racism, and German National Socialism* (New York: Oxford University Press, 2002), 26.

4. "Biological Purge," *New York Times*, February 21, 1937.

5. I. L. Baldwin, "Chemicals and Pests," *Science*, September 28, 1962, 1042–1043.

6. Paul R. Ehrlich, Anne H. Ehrlich, and John P. Holdren, *Ecoscience: Population, Resources, Environment*, 3rd ed. (San Francisco: W. H. Freeman, 1978), 786–787, 837.

7. L. M. McKenzie, R. Z. Witter, L. S. Newman, and J. L. Adgate, "Human Health Risk Assessment of Air Emissions from Development of Unconventional Natural Gas Resources," *Science of the Total Environment* 424 (May 1, 2012): 79–87.

Chapter 2

1. Gardiner Harris and William Broad, "Scientists Welcome Obama's Words," *New York Times*, January 21, 2009, http://www.nytimes.com/2009/01/22/us/politics/22 science.html.

2. National Institutes of Health, "Stem Cell Information: Executive Summary," accessed January 19, 2012, http://stemcells.nih.gov/info/scireport/execSum.asp.

3. Jesse Derris, "Life Support? Stem-Cell Backing Holds at Six in 10," ABC News, August 3, 2001, http://abcnews.go.com/sections/politics/DailyNews/poll010803.html.

4. Amanda Gardner, "Most Americans Back Embryonic Stem Cell Research: Poll," *U.S. News & World Report*, October 7, 2010, http://health.usnews.com/health -news/managing-your-healthcare/research/articles/2010/10/07/most-americans-back -embryonic-stem-cell-research-poll.

5. Rachel Benson Gold, "Embryonic Stem Cell Research—Old Controversy; New Debate," *Guttmacher Report on Public Policy* 7, no. 4 (October 2004), http://www.guttmacher.org/pubs/tgr/07/4/gr070404.html.

6. National Institutes of Health, "NIH Human Embryonic Stem Cell Registry Under Former President Bush (August 9, 2001–March 9, 2009)," March 9, 2010, http://stemcells.nih.gov/research/registry/eligibilitycriteria.asp.

7. National Institutes of Health, "NIH Statement on the President's Stem Cell Address," August 9, 2001, http://www.nih.gov/news/pr/aug2001/od-09.htm.

8. David Jackson, "Bush Cites 'Moral Line' in Second Veto of Stem Cell Bill," *USA Today*, June 21, 2007, http://www.usatoday.com/news/washington/2007–06–20-bush-stem-cells_N.htm.

9. U.S. Department of Health and Human Services, "NIH Human Embryonic Stem Cell Registry," accessed January 3, 2012, http://grants.nih.gov/stem_cells/registry /current.htm.

10. Meredith Wadman, "Draft Stem Cell Guidelines Criticized," *Nature News*, May 27, 2009, http://www.nature.com/news/2009/090526/full/news.2009.515.html.

11. National Institutes of Health, "National Institutes of Health Guidelines on Human Stem Cell Research," January 12, 2011, http://stemcells.nih.gov/policy /2009guidelines.htm.

12. Steve Connor, "Scientific Illiteracy All the Rage Among the Glitterati," *The Independent*, December 27, 2008, http://www.independent.co.uk/news/science/scientific-illiteracy-all-the-rage-among-the-glitterati-1212406.html.

13. Kreesten Madsen et al., "A Population-Based Study of Measles, Mumps, and Rubella Vaccination and Autism," *New England Journal of Medicine* 347 (2002): 1477–1482, http://www.nejm.org/doi/full/10.1056/NEJMoa021134.

14. Scott Gottlieb, "Why You Can't Get the Swine Flu Vaccine," *Wall Street Journal*, October 27, 2009, http://online.wsj.com/article/SB100014240527487043359045744 97324151841690.html.

15. Ibid.

16. Andrew Pollack, "Benefit and Doubt in Vaccine Additive," *New York Times*, September 21, 2009, http://www.nytimes.com/2009/09/22/health/22vacc.html?page wanted=all.

17. Bill Frezza, "Junk Science Returns to the White House," RealClearMarkets, November 2, 2009, http://www.realclearmarkets.com/articles/2009/11/02/junk _science_returns_to_the_white_house_97481.html.

18. Centers for Disease Control and Prevention, "Updated CDC Estimates of 2009 H1N1 Influenza Cases, Hospitalizations, and Deaths in the United States, April 2009–April 10, 2010," May 14, 2010, http://www.cdc.gov/h1n1flu/estimates_2009 _h1n1.htm.

19. Christopher Marquis, "Bush Misuses Science Data, Report Says," *New York Times*, August 8, 2003, http://www.nytimes.com/2003/08/08/politics/08REPO.html.

20. White House Press Office, "Remarks of President Barack Obama—as Prepared for Delivery: Signing of Stem Cell Executive Order and Scientific Integrity Presidential Memorandum," March 9, 2009, http://www.whitehouse.gov/the_press_office /Remarks-of-the-President-As-Prepared-for-Delivery-Signing-of-Stem-Cell-Executive -Order-and-Scientific-Integrity-Presidential-Memorandum.

21. Neela Banerjee, "Report Critical of Government Response to Gulf Oil Spill," *Los Angeles Times*, October 6, 2010, http://articles.latimes.com/2010/oct/06/nation /la-na-1007-oil-spill-criticism-20101007.

22. Dan Berman, "Interior Inspector General: White House Skewed Drilling-Ban Report," Politico, November 9, 2010, http://www.politico.com/news/stories/1110/ 44921.html.

23. Riccardo Polosa et al., "Effect of an Electronic Nicotine Delivery Device (e-Cigarette) on Smoking Reduction and Cessation: A Prospective 6-Month Pilot Study," *BMC Public Health* 11 (2011): 786.

24. John Tierney, "A Tool to Quit Smoking Has Some Unlikely Critics," *New York Times*, November 7, 2011, http://www.nytimes.com/2011/11/08/science/e-cigarettes -help-smokers-quit-but-they-have-some-unlikely-critics.html.

25. John Stossel, "The FDA Kills Smokers," Reason, November 17, 2011, http://reason.com/archives/2011/11/17/the-fda-kills-smokers.

26. "FDA Limits Some Antibiotics in Livestock," *USA Today*, January 4, 2012, http://yourlife.usatoday.com/health/story/2012–01–04/FDA-limits-some-antibiotics -in-livestock/52380100/1.

27. Food and Drug Administration, "2010 Summary Report on Antimicrobials Sold or Distributed for Use in Food-Producing Animals," 2011.

28. See "FDA Limits Some Antibiotics in Livestock."

29. Robert Bryce, "If Reid, Obama Kill Yucca Mountain, Where Will Nuclear Waste Go? Think Fusion," *U.S. News & World Report*, June 24, 2009, http://www .usnews.com/opinion/articles/2009/06/24/if-reid-obama-kill-yucca-mountain-where- will-nuclear-waste-go-think-fusion.

30. Matthew Wald, "Spent Fuel Pools as a Bright Spot in Fukushima's Crisis," *New York Times*, July 26, 2011, http://green.blogs.nytimes.com/2011/07/26/spent -fuel-pools-in-japan-survived-disaster-n-r-c-report-says/.

31. Dawn Stover, "The 'Scientization' of Yucca Mountain," *Bulletin of the Atomic Scientists*, October 12, 2011, http://www.thebulletin.org/web-edition/columnists/ dawn-stover/the-scientization-of-yucca-mountain.

32. Eliot Marshall, "Republicans Charge 'Impropriety' in Halting Yucca Mountain Safety Review," *Science Insider*, October 14, 2010, http://news.sciencemag.org /scienceinsider/2010/10/republicans-charge-impropriety.html.

33. Stephen Power, "Report Slams U.S. Nuclear Regulator," *Wall Street Journal*, June 10, 2011, http://online.wsj.com/article/SB10001424052702304259304576375 961521636474.html.

34. See Stover, "The 'Scientization' of Yucca Mountain."

35. Kevin Bullis, "Q&A: Steven Chu," *MIT Technology Review*, May 14, 2009, http://www.technologyreview.com/business/22651/.

36. Ibid.

37. "Cash for Clunkers Results Finally In: Taxpayers Paid $24,000 per Vehicle Sold," Edmunds.com, October 29, 2009, http://www.edmunds.com/about/press/cash -for-clunkers-results-finally-in-taxpayers-paid-24000-per-vehicle-sold-reports -edmundscom.html.

38. Charles Herman, "'Cash for Clunkers' Environmental Benefits Are in Doubt," ABC News, August 24, 2009, http://abcnews.go.com/Business/JustOneThing/cash -clunkers-environmental-benefits-doubt/story?id=8386275.

39. Emily Badger, "Cash for Clunkers Was a Clunker," Miller-McCune, November 2, 2011, http://www.miller-mccune.com/business-economics/cash-for-clunkers-was -a-clunker-37532/.

40. Ibid.

41. "Cash for Clunkers Results Finally In."

42. Ben Geman and Erik Wasson, "Obama Shelves EPA Smog Rule in Huge Defeat for Environmental Groups," *The Hill*, September 2, 2011, http://thehill.com/blogs/e2 -wire/e2-wire/179357-white-house-shelves-smog-rule-in-huge-defeat-for-green-groups.

43. "Biofools," *The Economist*, April 8, 2009, http://www.economist.com/ node/13437705.

44. Larry Rohter, "Obama Camp Closely Linked with Ethanol," *New York Times*, June 23, 2008, http://www.nytimes.com/2008/06/23/us/politics/23ethanol.html.

45. Jonathan Weisman, "Romney Hearts Ethanol Subsidies," *Wall Street Journal*, May 27, 2011, http://blogs.wsj.com/washwire/2011/05/27/romney-hearts-ethanol -subsidies/.

46. Gerard Wynn, "U.S. Corn Ethanol 'Was Not a Good Policy'—Gore," Reuters, November 22, 2010, http://af.reuters.com/article/energyOilNews/idAFLDE6AL0 YT20101122.

47. Miguel Llanos, "$6-Billion-a-Year Ethanol Subsidy Dies—but Wait There's More," MSNBC, December 29, 2011, http://usnews.msnbc.msn.com/_news/2011/ 12/29/9804028–6-billion-a-year-ethanol-subsidy-dies-but-wait-theres-more.

48. "Barack Obama's Remarks in St. Paul," *New York Times*, June 3, 2008, http://www.nytimes.com/2008/06/03/us/politics/03text-obama.html.

49. Felicity Barringer, "For New Generation of Power Plants, a New Emission Rule from the E.P.A.," *New York Times*, March 27, 2012, http://www.nytimes.com/2012/ 03/28/science/earth/epa-sets-greenhouse-emission-limits-on-new-power-plants.html.

50. American Clean Energy and Security Act of 2009, H.R. 2454, 111th Congress (2009–2010), http://thomas.loc.gov/cgi-bin/bdquery/z?d111:H.R.2454.

51. "WV-Sen: Joe Manchin Shoots Cap and Trade Bill," RealClearPolitics, October 11, 2010, http://www.realclearpolitics.com/video/2010/10/11/wv-sen_manchin_ad_ dead_aim.html.

Chapter 3

1. Greg Miller, "Why the 'Prius Driving, Composting' Set Fears Vaccines," *Science Insider*, January 31, 2011, http://news.sciencemag.org/scienceinsider/ 2011/01/why-the-prius-driving-composting.html?ref=hp.

2. David Wasserman, "Will the 2012 Election Be a Contest of Whole Foods vs. Cracker Barrel Shoppers?," *Washington Post*, December 9, 2011, http://www.washingtonpost.com/opinions/will-the-2012-election-be-a-contest-of-

whole-foods-vs-cracker-barrel-shoppers/2011/09/28/gIQAMuXDiO_story.html.

3. Nick Paumgarten, "Food Fighter," *New Yorker,* January 4, 2010, http://www.newyorker.com/reporting/2010/01/04/100104fa_fact_paumgarten.

4. Rebecca Boyle, "Atomic Gardens, the Biotechnology of the Past, Can Teach Lessons About the Future of Farming," *Popular Science*, April 22, 2011, http://www.popsci.com/technology/article/2011–04/atomic-gardens-biotech nology-past-can-teach-lessons-about-future-farming.

5. David Clark and Nanette Pazdernik, *Biotechnology: Applying the Genetic Revolution* (Burlington, MA: Elsevier Academic Press, 2009), 413.

6. Dan Charles, "Grainy Season: Engineering Drought-Resistant Wheat," NPR, October 22, 2010, http://www.npr.org/templates/story/story.php?storyId=13 0734134.

7. "Golden Rice an Effective Source of Vitamin A," *Science Daily*, May 18, 2009, http://www.sciencedaily.com/releases/2009/05/090513121102.htm.

8. Xan Rice, "Ugandan Scientists Grow GM Banana as Disease Threatens Country's Staple Food," *The Guardian*, March 9, 2011 http://www.guardian .co.uk/world/2011/mar/09/gm-banana-crop-disease-uganda.

9. Tiffany Kaiser, "Researchers Fight Global Warming with Genetically Altered Trees," *Daily Tech*, October 1, 2010, http://www.dailytech.com/Researchers +Fight+Global+Warming+with+Genetically+Altered+Trees/article19776.htm.

10. Marin Enserink, "'Disease-Proof Mosquitoes' Could Spread Like Wild-fire," *Science Now*, April 20, 2011, http://news.sciencemag.org/sciencenow /2011/04/disease-proof-mosquito-could-spr.html?ref=hp.

11. Steve Connor, "GM Lab Creates Chicken That Cannot Spread Bird Flu," *The Independent*, January 14, 2011, http://www.independent.co.uk/news /science/gm-lab-creates-chicken-that-cannot-spread-bird-flu-2184280.html.

12. Linda Bren, "Genetic Engineering: The Future of Foods?," *FDA Consumer Magazine*, November 2003, http://permanent.access.gpo.gov/lps1609/www.fda .gov/fdac/features/2003/603_food.html.

13. Jennifer Carpenter, "GM Crops Can Benefit Farmers," *The Guardian*, April 21, 2010, http://www.guardian.co.uk/commentisfree/cif-green/2010/apr/ 21/gm-crops-benefit-farmers.

14. Ibid.

15. Brian Palmer, "How Can the FDA Tell Whether GMO Fish Are Safe?," *Slate*, September 21, 2010, http://www.slate.com/id/2268085/.

16. James H. Maryanski, Congressional testimony, Food and Drug Adminis-tration, October 19, 1999, http://www.fda.gov/newsevents/testimony/ucm115 032.htm.

17. See http://www.glofish.com/.

18. Larry Bell, "Climate Change as Religion: The Gospel According to Al Gore," *Forbes*, April 26, 2011, http://blogs.forbes.com/larrybell/2011/04/26 /climate-change-as-religion-the-gospel-according-to-gore/?partner=contextstory.

19. "Organic Food," *Penn & Teller: Bullshit!*, July 30, 2009.

20. "Get Your Deuces Flowing with Strawberries This Summer," *Which?*, June 21, 2008, http://www.which.co.uk/about-which/press/press-releases/product-press-releases/which-magazine/2008/06/get-your-deuces-flowing-with-strawber-ries-this-summer/.

21. Emily Sohn, "Organic Veggies Not More Nutritious," *Discovery News*, November 22, 2010, http://news.discovery.com/human/organic-vegetables-an-tioxidants.html.

22. Candy Sagon, "Grass-Fed Beef Called Healthier," *Washington Post*, March

15, 2006, http://www.washingtonpost.com/wp-dyn/content/article/2006/03/14/AR2006031400312.html.

23. C. S. Curtis and S. Misner, "Pesticide Versus Organically Grown Food," University of Arizona, 2006, http://ag.arizona.edu/pubs/health/foodsafety/az1079.html.

24. "Organic Pesticides Not Always 'Greener' Choice, Study Finds," *Science Daily*, June 23, 2010, http://www.sciencedaily.com/releases/2010/06/100622175510.htm.

25. Banister Nuri, "Organic Farming Is More Efficient Than Conventional," *Journal of Young Investigators*, July 5, 2005, http://www.jyi.org/news/nb.php?id=380.

26. Ronald Bailey, "Organic Alchemy," Reason, June 5, 2002, http://reason.com/archives/2002/06/05/organic-alchemy.

27. Brain Palmer, "Vinho Verde," *Slate*, March 22, 2011, http://www.slate.com/id/2288521/.

28. "Organic Food Miles Take Toll on Environment," *Science Daily*, June 7, 2007, http://www.sciencedaily.com/releases/2007/06/070606113311.htm.

29. Mischa Popoff, "We Must End the Organic Food Scam Now," RealClear-Science, September 5, 2011, http://www.realclearscience.com/articles/2011/09/05/we_must_end_the_organic_food_scam_now_106251.html.

30. Consumers Union, "National Organic Program," April 10, 1998, http://www.consumersunion.org/pub/core_food_safety/002315.html.

31. Li Ping, "Western Countries Say 'No' to Chinese Organic Food," *Epoch Times*, July 25, 2011, http://www.theepochtimes.com/n2/china-news/chinese-organic-exports-wear-out-their-welcome-59531.html.

32. William Neuman and David Barboza, "U.S. Drops Inspector of Food in China," *New York Times*, June 13, 2010, http://www.nytimes.com/2010/06/14/business/global/14organic.html.

33. Mischa Popoff, personal communication, November 8, 2011.

34. Mischa Popoff, "How Organic Activists Spread Misinformation," Real-ClearScience, October 3, 2011, http://www.realclearscience.com/articles/2011/10/03/how_organic_activists_spread_misinformation_106252.html.

35. Carl Winter and Josh Katz, "Dietary Exposure to Pesticide Residues from Commodities Alleged to Contain the Highest Contamination Levels," *Journal of Toxicology* 2011 (2011): 1–7, http://www.ncbi.nlm.nih.gov/pmc/articles/PMC3135239/.

36. Bauman Research Group at Cornell University, "rBST Information Booklet," August 31, 2009, http://www.ansci.cornell.edu/bauman/envir_impact/rbst_booklet.html.

37. Food and Drug Administration, "Report on the Food and Drug Administration's Review of the Safety of Recombinant Bovine Somatotropin," April 23, 2009, http://www.fda.gov/AnimalVeterinary/SafetyHealth/ProductSafetyInformation/ucm130321.htm.

38. Ibid.

39. Ibid.

40. See Bauman Research Group, "rBST Information Booklet."

41. Dan Flynn, "Odwalla Apple Juice *E. coli* Outbreak," *Food Safety News*, September 17, 2009, http://www.foodsafetynews.com/2009/09/meaningful-outbreak-4-odwalla-apple-juice-e-coli-o157h7-outbreak/.

42. Maureen O'Hagan, "Is Raw, Unpasteurized Milk Safe?," *Seattle Times*, March 20, 2010, http://seattletimes.nwsource.com/html/localnews/2011399591_rawmilk21m.html.

43. Jeffrey LeJeune and Paivi Rajala-Schultz, "Unpasteurized Milk: A Continued Public Health Threat," *Clinical Infectious Disease* 48, no. 1 (2009): 93–100, http://cid.oxfordjournals.org/content/48/1/93.full.

44. Laura Landro, "A Clash over Unpasteurized Milk Gets Raw," *Wall Street Journal*, March 30, 2010, http://online.wsj.com/article/SB1000142405270230 4370304575151663770115120.html.

45. Richard Raymond, "There Ought to Be a Milk Law," *Food Safety News*, August 18, 2011,

Chapter 4

1. Alternative Fuels and Advanced Vehicles Data Center (AFDC), "Properties of Fuels," U.S. Department of Energy, accessed July 27, 2007, http://www.afdc .energy.gov/afdc/pdfs/fueltable.pdf.

2. Food and Agriculture Organization, "Livestock's Long Shadow," 2006, ftp://ftp.fao.org/docrep/fao/010/a0701e/a0701e.pdf.

3. Chris Goodall, "How to Live a Low-Carbon Life," *NewScientist*, April 21, 2007.

4. Annabel L. Merrill and Bernice K. Watt, "Energy Value of Foods: Basis and Derivation," Agriculture Handbook No. 74 (Washington, DC: U.S. Department of Agriculture, February 1973), http://www.nal.usda.gov/fnic/foodcomp/Data/ Classics?ah74.pdf.

5. "How many calories in a pound of beef?," Wiki.answers.com, http://wiki .answers.com/Q/How_many_calories_in_a_pound_of_ground_beef.

6. Andrew C. Kimball, "On the Road: Steering Towards Ecological Disaster," in *The Green Lifestyle Handbook*, ed. Jeremy Rifkin (New York: Henry Holt, 1990), 33–42.

7. Alan B. Durning, "Cost of Beef for Health and Habitat," *Los Angeles Times*, September 21, 1986, http://articles.latimes.com/1986–09–21/opinion/op-9037_1 _meat-consumption.

8. Nathan Fiala, "How Meat Contributes to Global Warming," *Scientific American*, February 4, 2009, http://www.scientificamerican.com/article.cfm?id _the-greenhouse-hamburger.

9. David Pimentel, "Livestock Production: Energy Inputs and the Environment," August 1997, http://www.news.cornell.edu/releases/aug97/livestock.hrs.html.

10. Paul R. Ehrlich, Anne H. Ehrlich, and John P. Holdren, *Ecoscience: Population, Resources, Environment*, 3rd ed. (San Francisco: W. H. Freeman, 1978), 942–943.

11. Bob McBride, "Broken Heartland: Farm Crisis in the Midwest," *The Nation*, February 8, 1986, 132–133.

12. David A. Stockman, *The Triumph of Politics: How the Reagan Revolution Failed* (New York: Harper and Row, 1986), 152–154.

13. Jesse H. Ausubel and Paul E. Waggoner, "Dematerialization: Variety, Caution, and Persistence," *Proceedings of the National Academy of Sciences* 105, no. 35 (2008): 12774–12779.

14. Ibid.

15. Jesse H. Ausubel, "The Great Reversal: Nature's Chance to Restore Land and Sea," Program for the Human Environment, Rockefeller University, http://phe.rockefeller.edu/great_reversal/.

16. P. E. Waggoner and J. H. Ausubel, "A Framework for Sustainability Science: A Renovated IPAT Identity," *Proceedings of the National Academy of*

Sciences 99, no. 12 (2002): 7860–7865, doi:10.1073/pnas.122235999.

17. Ron Bailey, "Deconsumption Versus Dematerialization," Reason.com, February 2011, http://reason.com/archives/2011/02/15/deconsumption-versus -demateria.

18. Mike Archer, "Ordering the Vegetarian Meal? There's More Animal Blood on Your Hands," *The Conversation*, December 16, 2011.

19. Ibid.

20. Hanna L. Tuomisto and M. Joost Teixeira de Mattos, "Environmental Impacts of Cultured Meat Production," *Environmental Science and Technology* 45, no. 14 (2011): 6117–6123, doi:10.1021/es200130u.

21. "The 50 Best Inventions of 2009," *Time* November 12, 2009, http://www.time.com/time/specials/packages/0,28757,1934027,00.html.

22. Winston Churchill, *Thoughts and Adventures* (London: Thornton Butterworth, Limited, 1932), 291.

23. Mark Rosegrant, Michael Paisner, Siet Meijer, and Julie Witcover, "2020 Global Food Outlook: Trends, Alternatives, and Choices" (Washington, DC: International Food Policy Research Institute, 2001).

24. Tuomisto and Teixeira de Mattos, "Environmental Impacts of Cultured Meat Production."

25. John Laumer, "Synthetic Meat Is a Ridiculous Delusion Offering No Environmental Benefit," Treehugger.com, November 15, 2011, http://www .treehugger.com/green-food/synthetic-meat-ridiculous-delusion-no-possible -market-appeal.html.

Chapter 5

1. Matthew Jaffe, "Rand Paul's Toilet Tirade," ABC News, March 10, 2011, http://abcnews.go.com/blogs/politics/2011/03/rand-pauls-toilet-tirade/.

2. J. F. Kenny et al., "Estimated Use of Water in the United States in 2005," U.S. Geological Survey Circular 1344, 2009, http://pubs.usgs.gov/circ/1344/.

3. Ibid.

4. Charles Fishman, "Five Myths About Water," *Washington Post*, April 6, 2012, http://www.washingtonpost.com/opinions/five-myths-about-water/2012/04 /06/gIQAS6EB0S_story_1.html.

5. Brad Lendon, "Low-Flow Toilets Cause Stink in San Francisco," CNN, March 1, 2011, http://news.blogs.cnn.com/2011/03/01/low-flow-toilets-cause-stink -in-san-francisco/.

6. Dan Sullivan, "Majority of Alaskans Agree with Drilling in ANWR," *U.S. News & World Report*, November 3, 2011, http://www.usnews.com/debate-club/is-it-time -to-drill-in-the-arctic-refuge/majority-of-alaskans-agree-with-drilling-in-anwr.

7. Todd Woody, "Battle Brewing over Giant Desert Solar Farm," *New York Times* "Green," August 5, 2009, http://green.blogs.nytimes.com/2009/08/05/battle-brewing -over-giant-desert-solar-farm/.

8. Felicity Barringer, "Environmentalists in a Clash of Goals," *New York Times*, March 23, 2009, http://www.nytimes.com/2009/03/24/science/earth/24ecowars.html.

9. See Woody, "Battle Brewing over Giant Desert Solar Farm."

10. Todd Woody, "A Move to Put the Union Label on Solar Power Plants," *New York Times*, June 18, 2009, http://www.nytimes.com/2009/06/19/business/energy -environment/19unions.html.

11. Scott Shackford, "'Greenmail' a Threat to Energy Projects," *Desert Dispatch*, February 14, 2011, http://www.desertdispatch.com/opinion/projects-10296-energy-calling.html.

12. Marc Lifsher, "Labor Coalition's Tactics on Renewable Energy Projects Are Criticized," *Los Angeles Times*, February 5, 2011, http://articles.latimes.com/2011/feb/05/business/la-fi-solar-unions-20110205.

13. Sara Mitchell, "Calico Solar Promises Construction Jobs to Unions," *Desert Dispatch*, August 14, 2011, http://www.desertdispatch.com/articles/solar-11394-calico-unions.html.

14. "California Unions for Reliable Energy Partners with K Road Calico Solar Project," *Business Wire*, August 5, 2011, http://www.businesswire.com/news/home/20110805006022/en/California-Unions-Reliable-Energy-Partners-Road-Calico.

15. Robert Farley, "Checking George Will on Birds and Wind Turbines," PolitiFact, May 3, 2010, http://www.politifact.com/truth-o-meter/article/2010/may/03/checking-george-will-birds-and-wind-turbines/.

16. Lisa Foderaro, "A City of Glass Towers, and a Hazard for Migratory Birds," *New York Times*, September 14, 2011, http://www.nytimes.com/2011/09/15/nyregion/making-new-yorks-glass-buildings-safer-for-birds.html?_r=2&ref=science.

17. Julia Layton, "Do Wind Turbines Kill Birds?," HowStuffWorks, accessed September 16, 2011, http://science.howstuffworks.com/environmental/green-science/wind-turbine-kill-birds.htm.

18. Ibid.

19. Kevin Drum, "Our National Cat Problem," *Mother Jones*, March 22, 2011, http://motherjones.com/kevin-drum/2011/03/our-national-cat-problem.

20. See Foderaro, "A City of Glass Towers."

21. Leslie Kaufman, "Conspiracies Don't Kill Birds. People, However, Do," *New York Times*, January 17, 2011, http://www.nytimes.com/2011/01/18/science/18birds.html.

22. Carla Hall, "Smithsonian Bird Researcher Is Convicted of Trying to Poison Cats," *Los Angeles Times*, November 1, 2011, http://opinion.latimes.com/opinionla/2011/11/a-dc-bird-researcher-is-convicted-of-trying-to-poison-cats.html.

23. Rome Neal, "Storm over Mass. Windmill Plan," CBS News, February 11, 2009, http://www.cbsnews.com/stories/2003/06/26/sunday/main560595.shtml.

24. Duncan Graham-Rowe, "Hydroelectric Power's Dirty Secret Revealed," *New-Scientist*, February 26, 2005, http://www.newscientist.com/article/mg18524884.100-hydroelectric-powers-dirty-secret-revealed.html.

25. "Greenhouse Gas Impact of Hydroelectric Reservoirs Downgraded," *Science Daily*, August 1, 2011, http://www.sciencedaily.com/releases/2011/08/110801134733.htm.

26. Ibid.

27. Michael Scott, "Hydroelectric Power Plant at Gorge Metro Dam Is Scrapped After EPA, Summit County Metro Parks, Environmentalists Complain," Cleveland.com, June 24, 2009, http://www.cleveland.com/science/index.ssf/2009/06/hydroelectric_power_plant_at_g.html.

28. Paulo Cabral, "Brazil Awards Rights to Develop Belo Monte Dam," BBC News, April 20, 2010, http://news.bbc.co.uk/2/hi/8633786.stm.

29. Russell McLendon, "What Are Rare Earth Metals?," Mother Nature Network, June 22, 2011, http://www.mnn.com/earth-matters/translating-uncle-sam/stories/what-are-rare-earth-metals.

30. David Biello, "Dark Side of Solar Cells Brightens," *Scientific American*, February 21, 2008, http://www.scientificamerican.com/article.cfm?id=solar-cells-prove -cleaner-way-to-produce-power.

31. Todd Myers, *Eco-Fads* (Seattle: Washington Policy Center, 2011), 11–14.

32. Susan Kelleher, "Seattle Refuses to Use Salt; Roads 'Snow-Packed' by Design," *Seattle Times*, December 23, 2008, http://seattletimes.nwsource.com/html/localnews/ 2008551284_snowcleanup23m.html.

33. Emily Heffter, "Seattle to Use Salt in Future Storms," *Seattle Times*, December 31, 2008, http://seattletimes.nwsource.com/html/localnews/2008577361_salt31m.html.

34. Todd Myers, interview, October 20, 2011.

35. See Kelleher, "Seattle Refuses to Use Salt."

36. Babbage, "The Difference Engine: The Really Big One," *The Economist*, March 18, 2011, http://www.economist.com/blogs/babbage/2011/03/megaquakes.

37. Washington State Department of Transportation, "Alaskan Way Viaduct and Seawall Replacement—Timeline," 2011, http://www.wsdot.wa.gov/Projects/Viaduct/ Schedule.htm.

38. Susannah Frame, "$400 Million Spent Before Cement Poured for New 520 Bridge," King 5 News, March 17, 2011,

http://www.king5.com/home/Investigators-520-bridge-118221934.html.

39. Mike Lindblom, "'Pacific Interchange' Plan for Highway 520 Gains Momentum," *Seattle Times*, September 29, 2006, http://seattletimes.nwsource.com/html/ localnews/2003280662_520bridge29m.html.

40. John Tierney, "Holdren's Ice Age Tidal Wave," *Tierney Lab*, September 29, 2009, http://tierneylab.blogs.nytimes.com/2009/09/29/dr-holdrens-ice-age-tidal-wave/.

41. Ibid.

42. S. Solomon et al., "Is Sea Level Rising?," in *IPCC Fourth Assessment Report* (New York: Oxford University Press, 2007), http://www.ipcc.ch/publications _and_data/ar4/wg1/en/faq-5–1.html.

43. Rob Young and Orrin Pilkey, "Get Ready for Seven-Foot Sea Level Rise as Climate Change Melts Ice Sheets," *The Guardian*, January 15, 2010, http://www .guardian.co.uk/environment/2010/jan/15/sea-level-climate-change.

44. Mike Dunford, "Did the President's Nominee for Science Advisor Say He Thinks a Billion People Are Gonna Die?," ScienceBlogs, February 13, 2009, http://scienceblogs.com/authority/2009/02/did_the_presidents_nominee_for.php.

45. David Harsanyi, "Science Fiction 'Czar,'" Reason, July 15, 2009, http://reason.com/archives/2009/07/15/science-fiction-czar.

46. "Glenn Beck Claims Science Czar John Holdren Proposed Forced Abortions and Putting Sterilants in the Drinking Water to Control Population," PolitiFact, July 29, 2009, http://www.politifact.com/truth-o-meter/statements/2009/jul/29/glenn-beck/glenn-beck-claims-science-czar-john-holdren-propos/.

47. Mary Ellen Harte and Anne Ehrlich, "The World's Biggest Problem? Too Many People," *Los Angeles Times*, July 21, 2011, http://articles.latimes.com/2011/ jul/21/opinion/la-oe-harte-population-20110721.

48. "The Seventh Billion," *The Economist*, November 22, 2010, http://www.economist.com/blogs/multimedia/2010/11/world_population.

49. "World Population in 2300," in *Proceedings of the United Nations Expert Meeting on World Population in 2300* (New York: United Nations, 2004), http://www.un.org/esa/population/publications/longrange2/2004worldpop2300 reportfinalc.pdf.

50. Fred Pearce, "When Will the 7 Billionth Human Be Born?," *NewScientist*, October 14, 2011, http://www.newscientist.com/article/mg21228344.500-when-will -the-7-billionth-human-be-born.html.

51. "Go Forth and Multiply a Lot Less," *The Economist*, October 29, 2009, http://www.economist.com/node/14743589.

52. "Low Fertility in Europe—Is It a Concern?," *Science Codex*, June 17, 2011, http://www.sciencecodex.com/low_fertility_in_europe_is_there_still_reason_to_worry.

53. Anya Ardayeva, "Demographers Warn of Looming Population Crisis for Russia," Voice of America, March 21, 2011, http://www.voanews.com/english/news/europe/ Demographers-Warn-of-Looming-Population-Crisis-for-Russia-118382619.html.

54. See "Go Forth and Multiply a Lot Less."

55. "A Tale of Three Islands," *The Economist*, October 22, 2011, http://www .economist.com/node/21533364.

56. Ibid.

57. "Gendercide," *The Economist*, March 4, 2010, http://www.economist.com/ node/15606229.

58. Vivek Prakash, "A Village of Eternal Bachelors," Reuters, October 11, 2011, http://blogs.reuters.com/photographers-blog/2011/10/12/a-village-of-eternal -bachelors/.

59. Ibid.

60. Alex Berezow, "Should We Feed Hungry People, Even if It's Bad for the Environment?," Forbes, April 5, 2011, http://www.forbes.com/2011/04/05/environment -hunger-population-opinions-alex-berezow.html.

Chapter 6

1. Brad Johnson, "Murkowski Wants to Save Alaska by Destroying It," Grist.org, March 5, 2010, http://grist.org/politics/2010–03–04-murkowski-wants-to-save-alaska -by-destroying-it/.

2. Sarah Hodgdon, "The Arctic Refuge Is Ready for a Real Commitment," Treehugger.com, November 18, 2010, http://www.treehugger.com/culture/the-arctic-refuge -is-ready-for-a-real-commitment.html.

3. ANWR Assessment Team, "Oil and Gas Resource Potential of the Arctic National Wildlife Refuge 1002 Area, Alaska," U.S. Geological Survey Open File Report 98–34, http://pubs.usgs.gov/of/1998/ofr-98–0034/.

4. Energy Information Administration, Office of Oil and Gas, "Overview of U.S. Legislation and Regulations Affecting Offshore Natural Gas and Oil Activity," U.S. Department of Energy, September 2005.

5. Doc Hastings, "Forget 10 Years—Drilling ANWR Would Pay Off Right Away," *U.S. News & World Report*, November 3, 2011, http://www.usnews.com/debate-club/is-it-time-to-drill-in-the-arctic-refuge/forget-10-years—drilling-anwr-would-pay -off-right-away.

6. Ibid.

7. Sean Cockeramham, "Senate Rejects Plan to Open Arctic Refuge to Drilling," *Sacramento Bee*, March 12, 2012.

8. "Energy Plays Key Role in Obama Speech," CNN Money, January 24, 2012.

9. Matthew Kennard, "US Greenhouse Gases Drop to 15-Year Low," *Financial Times*, April 18, 2011, http://www.ft.com/intl/cms/s/0/dfc8a384–6a06–11e0–86e4 –00144feab49a.html#axzz1rro81vaN.

10. Ibid.

11. Adam J. Liska et al., "Improvements in Life Cycle Energy Efficiency and Greenhouse Gas Emissions of Corn-Ethanol," *Journal of Industrial Ecology* 13, no. 1 (February 2009): 58–74, doi:10.1111/j.1530–9290.2008.00105.

12. "Biofools," *The Economist*, April 28, 2009.

13. "U.S. Corn Ethanol 'Was Not a Good Policy'—Gore," Reuters, November 22, 2010. http://www.reuters.com/article/2010/11/22/us-ethanol-gore-idUSTRE6AL3CN 20101122.

14. Peter A. Coates, "The Trans-Alaskan Pipeline Controversy: Technology, Conservation, and the Frontier," University of Alaska–Fairbanks, press release, October 1993, 196.

15. Thomas M. Brown, "That Unstoppable Pipeline; Our Arctic Will Never Be the Same," *New York Times*, October 14, 1973.

16. U.S. Fish and Wildlife Service, "Potential Impacts of Proposed Oil and Gas Development on the Arctic Refuge's Coastal Plain: Historical Overview and Issues of Concern," January 17, 2001, http://arctic.fws.gov/issues1.htm.

17. U.S. Department of State, "Keystone XL Project Final Environmental Impact Statement," August 26, 2011; and Institute for Energy Research, "US Energy Facts."

18. "The Truth About Aquifers," *New York Times*, October 4, 2011.

19. Perryman Group, "The Impact of Developing the Keystone XL Pipeline Project on Business Activity in the US," June 2010.

20. James Goeke, "The Truth About Aquifers," *New York Times*, October 4, 2011.

21. "Chu Says U.S. Energy Security 'Trade Off' Favors Oil-Sands Pipeline," *EnergyNow!*, August 30, 2011, video interview, http://www.youtube.com/watch?v= cjQm4hVwWkA.

22. "Keystone XL: Canada-US Pipeline Route Decision Delayed," BBC News, November 10, 2011.

23. Roger Runningen and Kate Andersen Brower, "Obama Says He Opposes Linking Keystone to Payroll Tax Cut," *Business Week*, December 13, 2011, http://www.businessweek.com/news/2011–12–13/obama-says-he-opposes-linking-keystone-to-payroll-tax-cut.html.

24. Rich Trzupek, "The Map Doesn't Lie: Keystone XL Pipeline Is Environmentally Safe," January 2012, Heartland.org.

25. Nebraska Keystone XL Facts page, http://www.keystonexlnebraska.com /the-facts/the-ogallala-aquifer.

26. Nebraska Pipeline Association home page, http://nebraskapipeline.com.

27. U.S. Department of State, "Keystone XL Project Final Environmental Impact Statement," August 26, 2011; and Institute for Energy Research, "US Energy Facts."

28. Trevor Butterworth, "Mining the Middle Ground," *The Daily*, April 23, 2012, http://www.thedaily.com/page/2012/04/23/042312-opinions-column-fracking-butterworth-1–2.

29. U.S. Department of State, "Keystone XL Pipeline Project Executive Summary of the Final EIS," August 26, 2011, http://keystonepipeline-xl.state.gov/documents /organization/181185.pdf.

30. U.S. Department of Energy, "The Recovery Act: Transforming America's Transportation Sector Batteries and Electric Vehicles," July 14, 2010.

31. Bill Moore, "Plug-In Hybrids: What's the Big Deal?," EVWorld.com, February 13, 2007, http://www.evworld.com/syndicated/evworld_article_1191.cfm.

32. General Motors, "Chevrolet Volt's Revolutionary Voltec Electric Drive System Delivers Efficiency with Performance," press release, October 10, 2010.

33. "Behind the Hidden Costs of Hybrids," Hybridcars.com, September 28, 2006, http://www.hybridcars.com/economics/hidden-costs.html.

34. Battery University at Cadex Electronics, "Nickel-Based Batteries," February 15, 2011, http://batteryuniversity.com/learn/article/Nickel_based_batteries.

35. Guillaume Majeau-Bettez, Troy R. Hawkins, and Anders Hammer Strømman, "Life Cycle Environmental Assessment of Lithium-Ion and Nickel Metal Hydride Batteries for Plug-In Hybrid and Battery Electric Vehicles," *Environmental Science and Technology* 45, no. 10 (2011): 4548–4554, doi:10.1021/es103607c.

36. U.S. Environmental Protection Agency, Air Pollution Control Orientation Course, January 29, 2010, http://www.epa.gov/apti/course422/.

37. Frank Didik, "Questions and Answers Regarding Electric Vehicles." Didik.com, 2005, http://www.didik.com/criticalev.htm.

38. Květoslava Burda and Marian Smoluchowski, "Why Is Photosynthesis Interesing?," *Foton* 93 (Spring, 2006), http://www.if.uj.edu.pl/Foton/92-special%20issue/pdf/06%20kburda.pdf.

39. U.S. Department of Energy, "Fuel Economy: Where the Energy Goes," http://www.fueleconomy.gov/feg/atv.shtml.

40. Tom Murphy, "Don't Be a PV Efficiency Snob," Do the Math, September 21, 2011, http://physics.ucsd.edu/do-the-math/2011/09/dont-be-a-pv-efficiency-snob/.

41. Eamon Javers, "DOE Pushing on with $5 Billion in Solar Energy Loans," CNBC, September 29, 2011, http://www.cnbc.com/id/44723203.

42. New Energy Finance, "Global Clean Energy Investment Overview," September 20, 2006, http://www.newenergyfinance.com/WhitePapers/download/6.

43. Juliet Eilperin, "Why the Clean Tech Boom Went Bust," *Wired*, January 20, 2012, http://www.wired.com/magazine/2012/01/ff_solyndra/all/1.

44. Ibid.

45. Rich Lowry, "Obama's Bad Bet on Green Energy," RealClearPolitics, September 2, 2011, http://www.realclearpolitics.com/articles/2011/09/02/obamas_bad_bet_on_green_energy_111184.html.

46. "CBS News: 11 More Solyndras in Obama Energy Program," RealClearPolitics, January 13, 2012, http://www.realclearpolitics.com/video/2012/01/13/cbs_news_11_more_solyndras_in_obama_energy_program.html.

47. Joe Stephens and Carol Leonnig, "Solyndra: Politics Infused Obama Energy Programs," *Washington Post*, December 25, 2011, http://www.washingtonpost.com/solyndra-politics-infused-obama-energy-programs/2011/12/14/gIQA4HllHP_story.html.

48. Susan Kraemer, "Obama Drives Clean Energy Boom on Public Lands," Earth Techling, January 10, 2012, http://www.earthtechling.com/2012/01/obama-drives-clean-energy-boom-on-public-lands/.

49. Energy Information Administration, "How Much Electricity Does an American Home Use?," U.S. Department of Energy, December 6, 2011.

50. Christiana Honsberg and Stuart Bowden, "Collection Probability," PV Education, http://www.pveducation.org/pvcdrom/solar-cell-operation/collection-probability.

51. "Power Grid of the Future Saves Energy," Fraunhofer Institute, November 2010, http://www.sciencecodex.com/power_grid_of_the_future_saves_energy.

52. Ibid.

53. S. 539: Clean Renewable Energy and Economic Development Act, 111th Congress, March 5, 2009, http://www.govtrack.us/congress/bills/111/s539/text.

54. Doug MacEachern, "Even with Solar, Power-Line Battles Loom," *Arizona Republic*, March 22, 2009.

55. Tom Murphy, "Don't Be a PV Efficiency Snob," Do the Math, September 21, 2011, http://physics.ucsd.edu/do-the-math/2011/09/dont-be-a-pv-efficiency-snob/,

56. Andrew C. Revkin, "Reactions to a New Plan for CO2 Progress," *New York Times*, September 27, 2011.

Chapter 7

1. Fiona Godlee, Jane Smith, and Harvey Marcovitch, "Wakefield's Article Linking MMR Vaccine and Autism Was Fraudulent," *British Medical Journal* 342 (2011): c7452, http://www.bmj.com/content/342/bmj.c7452.

2. Alice Park, "Great Science Frauds," *Time*, January 12, 2012, http://health land.time.com/2012/01/13/great-science-frauds/slide/andrew-wakefield/.

3. Ivan Oransky, "*Salon* Retracts 2005 Robert F. Kennedy, Jr. Piece on Alleged Autism-Vaccine Link," Retraction Watch, January 16, 2011, http://retractionwatch .wordpress.com/2011/01/16/salon-retracts-2005-robert-f-kennedy-jr-piece-on-alleged -autism-vaccine-link/.

4. Amy Wallace, "An Epidemic of Fear: How Panicked Parents Skipping Shots Endangers Us All," *Wired*, October 19, 2009, http://www.wired.com/magazine/2009/10/ff_waronscience/all/1.

5. Steve Krakauer, "Bill Maher's Medical Meltdown Turns Off Guests, Confuses Audience," Mediaite, October 18, 2009. http://www.mediaite.com/tv/bill-mahers -medical-meltdown-turns-off-guests-confuses-audience/.

6. Orac, "Bill Maher Gets the Richard Dawkins Award? That's Like Jenny McCarthy Getting an Award for Public Health," ScienceBlogs, July 23, 2009, http:// scienceblogs.com/insolence/2009/07/bill_maher_gets_the_richard_dawkins_awar.php.

7. Scott Hensley, "Pediatricians Fact-Check Bachmann's Bashing of HPV Vaccine," NPR, September 13, 2011, http://www.npr.org/blogs/health/2011/09/13/140445104 /pediatricians-fact-check-bachmanns-bashing-of-hpv-vaccine.

8. Greg Miller, "Why the 'Prius Driving, Composting' Set Fears Vaccines," *Science Insider*, January 31, 2011, http://news.sciencemag.org/scienceinsider/2011/01/why -the-prius-driving-composting.html?ref=hp.

9. Centers for Disease Control and Prevention, "Morbidity and Mortality Weekly Report," June 3, 2011, http://www.cdc.gov/mmwr/preview/mmwrhtml/mm6021a4 .htm?s_cid=mm6021a4_w.

10. Pauline Bartolone, "Opting Out: Community Immunity and the Choice to Vaccinate," *Capital Public Radio*, August 29, 2011, http://www.capradio.org/articles/ 2011/08/29/opting-out-community-immunity-and-the-choice-to-vaccinate.

11. Megan McArdle, "A Shocking Chart on Vaccination," *The Atlantic*, October 31, 2011, http://www.theatlantic.com/life/archive/2011/10/a-shocking-chart-on -vaccination/247651/.

12. Waldorf School of the Peninsula, "Steiner-Inspired Initiatives," accessed November 15, 2011, http://www.waldorfpeninsula.org/about/steiner_initiatives.html.

13. Michael Madigan, John Martinko, Paul Dunlap, and David Clark, *Brock Biology of Microorganisms* (San Francisco: Pearson Benjamin Cummings, 2009), 9.

14. Centers for Disease Control and Prevention, "Leading Causes of Death," accessed November 18, 2011, http://www.cdc.gov/nchs/fastats/lcod.htm.

15. Michael Specter, *Denialism* (New York: Penguin Press, 2009), 54.

16. Andrew Buncombe and Nina Lakhani, "Without Consent: How Drugs Companies Exploit Indian 'Guinea Pigs,'" *The Independent*, November 14, 2011, http://www.independent.co.uk/news/world/asia/without-consent-how-drugs-compa-nies-exploit-indian-guinea-pigs-6261919.html.

17. Ben Goldacre, "The True Purpose of a Drug Trial Is Not Always Obvious," *The Guardian*, July 1, 2011, http://www.guardian.co.uk/commentisfree/2011/jul/01/bad-science-drug-trials-seeding-trials.

18. Elliot Ross, "How Drug Companies' PR Tactics Skew the Presentation of Medical Reseach," *The Guardian*, May 20, 2011, http://www.guardian.co.uk/science/2011/may/20/drug-companies-ghost-writing-journalism.

19. Ibid.

20. Jacquelyn Smith, "America's Most Generous Companies," *Forbes*, October 28, 2010, http://www.forbes.com/2010/10/28/most-generous-companies-leadership-corporate-citizenship-philanthropy.html.

21. Eric Frazier and Marisa Lopez-Rivera, "Corporate Giving Slow to Recover as Economy Remains Shaky," *Chronicle of Philanthropy*, July 24, 2011, http://philanthropy.com/article/Big-Businesses-Won-t/128327/.

22. Susan Harding, "Health Plan Covers Assisted Suicide but Not New Cancer Treatment," KVAL, July 31, 2008, http://www.kval.com/news/26140519.html.

23. "Top Industries: Most Profitable," CNN Money, May 4, 2009, http://money.cnn.com/magazines/fortune/fortune500/2009/performers/industries/profits/.

24. Lara Salahi, "Obama Issues Executive Order to Ease Drug Shortages," ABC News, October 31, 2011, http://abcnews.go.com/Health/Wellness/presidents-executive-order-drug-shortage-draws-mixed-reactions/story?id=14852829#.Tsr6BnGkB7w.

25. Nidhi Subbaraman, "Flawed Arithmetic on Drug Development Costs," *Nature Biotechnology* 29 (2011): 381, http://www.nature.com/nbt/journal/v29/n5/full/nbt0511-381a.html.

26. Timothy Noah, "The Make-Believe Billion," *Slate*, March 3, 2011, http://www.slate.com/articles/business/the_customer/2011/03/the_makebelieve_billion.html.

27. Emily Willingham, "Huffington Post: Irresponsible Mouthpiece for the World of Woo," *Field of Science*, October 23, 2011, http://biologyfiles.fieldofscience.com/2011/10/huffington-post-irresponsible.html.

28. "Think Yourself Better," *The Economist*, May 19, 2011, http://www.economist.com/node/18710090.

29. Ibid.

30. "$34 Billion Spent Yearly on Alternative Medicine," MSNBC, July 30, 2009, http://www.msnbc.msn.com/id/32219873/ns/health-alternative_medicine/t/billion-spent-yearly-alternative-medicine/#.Ts2wEnGkB7w.

31. Leon Jaroff, "Bee Pollen Bureaucracy," *New York Times*, October 6, 1997, http://www.nytimes.com/1997/10/06/opinion/bee-pollen-bureaucracy.html.

32. "$2.5 Billion Spent, No Alternative Cures Found," MSNBC, June 10, 2009, http://www.msnbc.msn.com/id/31190909/ns/health-alternative_medicine/t/billion-spent-no-alternative-cures-found/.

33. Steven Salzberg, "The $350,000 Questionnaire from NIH," *Field of Science*, August 29, 2011, http://genome.fieldofscience.com/2011/08/350000-questionnaire-from-nih.html.

34. See "$2.5 Billion Spent."

35. See Specter, *Denialism*, 180.

36. David Gorski, "Why Senator Tom Harkin Should Be Considered a Public Health Menace," NCCAM Watch, May 3, 2011, http://www.nccamwatch.org/commentary/harkin.shtml.

37. See "Think Yourself Better."

38. Martin Enserink, "These Fake Pills May Help You Feel Better," *Science Now*, December 22, 2010, http://news.sciencemag.org/sciencenow/2010/12/these-fake-pills -may-help-you-fe.html?ref=hp.

39. Michael Berens and Christine Willmsen, "How One Man's Invention Is Part of a Growing Worldwide Scam That Snares the Desperately Ill," *Seattle Times*, November 19, 2007, http://seattletimes.nwsource.com/html/localnews/2004020583_miracle18 m2.html.

40. Ibid.

41. John Stossel, *Stossel*, April 1, 2010.

42. Deborah Blum, "A Chemical-Free Resolution," *PLoS Blogs*, January 5, 2011, http://blogs.plos.org/speakeasyscience/2011/01/05/a-chemical-free-resolution/.

43. Jon Entine, *Scared to Death: How Chemophobia Threatens Public Health* (New York: American Council on Science and Health, 2011), 26–27.

44. Jon Entine, "A Toxic Setback for the Anti-Plastic Campaigners," *The American*, April 19, 2011, http://www.american.com/archive/2011/april/a-toxic-setback-for -the-anti-plastic-campaigners.

Chapter 8

1. Joel Moreno, "Topless PETA Activists Protest Vets' Fish Toss," KOMO News, July 11, 2009, http://www.komonews.com/news/local/50546027.html.

2. Daniel Cressey, "Animal Research: Battle Scars," *Nature News*, February 23, 2011, http://www.nature.com/news/2011/110223/full/470452a.html.

3. P. Michael Conn and James Parker, "America's Other Most Wanted," *Wall Street Journal*, May 18, 2011, http://online.wsj.com/article/SB1000142405274870342120 4576327690420418996.html.

4. Oliver Wright, "Animal Rights Group Declares War on Leading Health Chari-ties," *The Independent*, June 21, 2011, http://www.independent.co.uk/news/uk/home -news/animal-rights-group-declares-war-on-leading-health-charities-2300281.html.

5. Ibid.

6. Ibid.

7. PETA, "Animal Rights Uncompromised: Life-Taking Charities," 2011, http://www.peta.org/about/why-peta/life-taking-charities.aspx.

8. Meredith Wadman, "Activists Ground Primate Flights," *Nature News*, March 20, 2012, http://www.nature.com/news/activists-ground-primate-flights-1.10255.

9. James McWilliams, "PETA's Terrible, Horrible, No Good, Very Bad History of Killing Animals," *The Atlantic*, March 12, 2012, http://www.theatlantic.com/health /archive/2012/03/petas-terrible-horrible-no-good-very-bad-history-of-killing-animals /254130/.

10. "Speaking Up: Confronting Misrepresentation," *Speaking of Research*, October 7, 2010, http://speakingofresearch.com/2010/10/07/speaking-up-confronting-misrep resentation/.

11. Ibid.

12. Center for Consumer Freedom, "Obama 'Regulatory Czar' Has Secret Animal-Rights Agenda, Says Consumer Group," press release, January 15, 2009, http://www .consumerfreedom.com/pressRelease_detail.cfm/r/249-obama-regulatory-czar-has -secret-animal-rights-agenda-says-consumer-group.

13. "Sunstein Has Said People Ought to Be Able to Sue on Behalf of Abused Ani-mals," PolitiFact, May 5, 2009, http://www.politifact.com/truth-o-meter/statements /2009/may/05/wayne-lapierre/sunstein-has-said-people-ought-be-able-sue-behalf-/.

14. See Conn and Parker, "America's Other Most Wanted."

15. Alok Jha, "One in Ten Research Projects Using Monkeys Has No Benefit," *The Guardian*, July 27, 2011, http://www.guardian.co.uk/science/2011/jul/27/research -projects-monkeys-benefit.

16. Animal Welfare Act, http://awic.nal.usda.gov/nal_display/index.php?info_cen- ter=3&tax_level=3&tax_subject=182&topic_id=1118&level3_id=6735&level4_id=0 &level5_id=0&placement_default=0.

17. American Psychological Association, "Rats, Mice, and Birds Excluded from Animal Welfare Act," July 2002, http://www.apa.org/monitor/julaug02/rats.aspx.

18. Meredith Wadman, "Animal Rights: Chimpanzee Research on Trial," *Nature News*, June 15, 2011, http://www.nature.com/news/2011/110615/full/474268a.html.

19. James Gorman, "U.S. Will Not Finance New Research on Chimps," *New York Times*, December 15, 2011, http://www.nytimes.com/2011/12/16/science/chimps -in-medical-research.html.

20. Richard Gray, "Drug and Cosmetics Firms Back Plan to Cut Animal Testing," *The Telegraph*, November 27, 2010, http://www.telegraph.co.uk/science/science -news/8165125/Drug-and-cosmetics-firms-back-plan-to-cut-animal-testing.html.

21. Lauran Neergard, "Government Adopts Strict Limits on Chimp Research," MSNBC, December 15, 2011, http://www.msnbc.msn.com/id/45684538/ns/health -health_care/t/government-adopts-strict-limits-chimp-research/#.T4yb3jI6X1Y.

22. D. L. Stokes, "Things We Like: Human Preferences Among Similar Organisms and Implications for Conservation," *Human Ecology* 35 (2007): 361–369, http://www.bothell.washington.edu/getattachment/IAS/faculty/dstokes/stokes-human- ecology.pdf.

23. Peter Singer, "Animal Liberation at 30," *New York Review of Books*, May 15, 2003, http://www.nybooks.com/articles/archives/2003/may/15/animal-liberation-at- 30/?pagination=false.

24. Peter Singer, "Heavy Petting," Utilitarian.net, 2001, http://www.utilitarian.net/ singer/by/2001——.htm.

25. Tony Fitzpatrick, "Biological Differences Among Races Do Not Exist, WU Re- search Shows," Washington University in St. Louis, October 15, 1998, http://wupa .wustl.edu/record_archive/1998/10–15–98/articles/races.html.

26. Ibid.

27. Lynn Jorde and Stephen Wooding, "Genetic Variation, Classification, and 'Race,'" *Nature Genetics* 36 (2004): S28–S33, http://www.nature.com/ng/journal /v36/n11s/full/ng1435.html.

28. Robyn Williams, "On Cloning, Genetic Selection, and Animal Rights" (inter- view with Peter Singer), Utilitarian.net, August 9, 2003, http://www.utilitarian.net/ singer/interviews-debates/20030809.htm.

29. Richard Corliss et al., "Should We All Be Vegetarians?," *Time*, July 15, 2002, http://www.time.com/time/magazine/article/0,9171,1002888–8,00.html.

30. Brian Palmer, "Would Your Dog Eat Your Dead Body?," *Slate*, July 13, 2011, http://www.slate.com/articles/news_and_politics/explainer/2011/07/would_your_dog _eat_your_dead_body.html.

31. "Killer Plant 'Eats' Great Tit at Somerset Nursery," BBC News, August 5, 2011, http://www.bbc.co.uk/news/uk-england-somerset-14416809.

32. James Gorman, "Animal Studies Cross Campus to Lecture Hall," *New York Times*, January 2, 2012, http://www.nytimes.com/2012/01/03/science/animal-studies- move-from-the-lab-to-the-lecture-hall.html?pagewanted=all.

33. "A Clarification on Canine Americans," NPR, September 19, 2006, http://www.npr.org/templates/story/story.php?storyId=6105346.

34. "Morrissey Stands by Norway Comments," *Toronto Sun*, July 30, 2011, http://www.torontosun.com/2011/07/30/morrissey-stands-by-norway-comments.

35. Marc Hauser, "The Origin of the Mind," *Scientific American*, September 2009, 44–51, http://www.scientificamerican.com/article.cfm?id=origin-of-the-mind.

36. Tom Jacobs, "A Chimp Couldn't Have Created That Painting," Miller-McCune, March 8, 2011, http://www.miller-mccune.com/culture-society/a-chimp-couldn%E2%80%99t-have-created-that-painting-28947/.

37. See Hauser, "The Origin of the Mind."

38. Stephanie Pappas, "Bird Brains: Pigeons Gamble Just like Humans," *Live Science*, October 13, 2010, http://www.livescience.com/8784-bird-brains-pigeons-gamble-humans.html.

39. John Timmer, "Bacteria Engage in Altruistic Suicide," *Ars Technica*, October 26, 2007, http://arstechnica.com/science/news/2007/10/bacteria-engage-in-altruistic-suicide.ars.

40. Ray Lilley, "Dolphin Saves Stuck Whales, Guides Them Back to Sea," *National Geographic News*, March 12, 2008, http://news.nationalgeographic.com/news/2008/03/080312-AP-dolph-whal.html.

41. Charles Q. Choi, "Selfless Chimps Shed Light on Evolution of Altruism," *Live Science*, June 25, 2007, http://www.livescience.com/4515-selfless-chimps-shed-light-evolution-altruism.html.

42. Dennis Miller, *Ranting Again* (New York: Doubleday, 1998), http://www.nytimes.com/books/first/m/miller-ranting.html.

43. Stephanie Pappas, "Humans on Verge of Causing 6th Great Mass Extinction," *Live Science*, March 2, 2011, http://www.livescience.com/13038-humans-causing-sixth-mass-extinction.html.

Chapter 9

1. European Science Foundation, "The Reality of Human Stem Cell Research in Europe," June 24, 2010, http://www.esf.org/media-centre/ext-single-news/article/the-reality-of-human-stem-cell-research-in-europe-621.html.

2. Frank Swain and Martin Robbins, "European Elections: The Anti-Science Sentiment Infecting Politics," *The Guardian*, June 1, 2009, http://www.guardian.co.uk/science/blog/2009/jun/01/european-elections-science-stem-cells-gm.

3. Jop de Vrieze, "First E.U. Science Adviser: 'We Are Sitting on a Goldmine,'" *Science Insider*, February 14, 2012, http://news.sciencemag.org/scienceinsider/2012/02/first-eu-science-adviser-we-are-.htm.

4. "Lessons from 'The Leopard,'" *The Economist*, December 10, 2009, http://www.economist.com/node/15065405.

5. Europa, "The Precautionary Principle," February 11, 2005, http://europa.eu/legislation_summaries/consumers/consumer_safety/l32042_en.htm.

6. Michael Specter, *Denialism* (New York: Penguin Press, 2009), 51–55.

7. "Swiss to Phase Out Nuclear Power," BBC, May 25, 2011, http://www.bbc.co.uk/news/world-europe-13549985.

8. "Germany: Nuclear Power Plants to Close by 2022," BBC, May 30, 2011, http://www.bbc.co.uk/news/world-europe-13592208.

9. "Italy Nuclear: Berlusconi Accepts Referendum Blow," BBC, June 14, 2011, http://www.bbc.co.uk/news/world-europe-13741105.

10. Josie Garthwaite, "Would a New Nuclear Plant Fare Better Than Fukushima?," *National Geographic*, March 23, 2011, http://news.nationalgeographic.com/news/energy/2011/03/110323-fukushima-japan-new-nuclear-plant-design/.

11. Alan Taylor, "Protesters Disrupt German Nuclear Waste Shipment," *The Atlantic*, November 28, 2011, http://www.theatlantic.com/infocus/2011/11/protesters-disrupt-german-nuclear-waste-shipment/100196/.

12. George Monbiot and Justin McCurry, "Post-Fukushima 'Anti-Radiation' Pills Condemned by Scientists," *The Guardian*, November 21, 2011, http://www.guardian.co.uk/environment/2011/nov/21/christopher-busby-radiation-pills-fukushima.

13. Ibid.

14. "Frack On," *The Economist*, November 26, 2011, http://www.economist.com/node/21540275.

15. Edward E. Cohen, Aubrey K. McClendon, and Paul Gallay, "The Battle over Fracking," *Wall Street Journal*, March 26, 2012,
http://online.wsj.com/article/SB10001424052702304724404577299662637393108.html.

16. "Russia Shuts Off Gas to Ukraine," BBC, January 1, 2009, http://news.bbc.co.uk/2/hi/europe/7806870.stm.

17. Geir Moulson, "*E. coli* Death Toll Up to at Least 47," PhysOrg.com, June 27, 2011, http://www.physorg.com/news/2011–06-coli-death-toll.html.

18. Thomas Friedman, *Longitudes and Attitudes* (New York: Anchor Books, 2003), 278–279.

19. Charlie Dunmore, "Biotech Firms Warn EU over Pace of GM Crop Approvals," Reuters, October 11, 2011, http://www.reuters.com/article/2011/10/11/us-eu-gmo-approvals-idUSTRE79A3G520111011.

20. Christie Wilcox, "Mythbusting 101: Organic Farming > Conventional Agriculture," *Scientific American*, July 18, 2011, http://blogs.scientificamerican.com/science-sushi/2011/07/18/mythbusting-101-organic-farming-conventional-agriculture/.

21. "Germany Bans Cultivation of GM Corn," *Der Spiegel*, April 14, 2009, http://www.spiegel.de/international/germany/0,1518,618913,00.html.

22. Barbara Casassus, "French Ban on Genetically Modified Corn Loses Another Round," *Science Insider*, November 28, 2011, http://news.sciencemag.org/science insider/2011/11/french-ban-on-genetically-modified.html?ref=hp.

23. Leigh Phillips, "EU Bans GM-Contaminated Honey from General Sale," *The Guardian*, September 7, 2011, http://www.guardian.co.uk/environment/2011/sep/07/europe-honey-gm.

24. David Derbyshire, "Europe's Opposition to GM Crops Is Arrogant Hypocrisy, Kenyan Scientist Warns," *The Guardian*, October 22, 2011, http://www.guardian.co.uk/environment/2011/oct/23/gm-crops-africa-biotechnology.

25. Council of Europe, "The Potential Dangers of Electromagnetic Fields and Their Effect on the Environment," May 6, 2011, http://assembly.coe.int/main.asp?Link=/documents/workingdocs/doc11/edoc12608.htm.

26. Michael Shermer, "Can You Hear Me Now? The Truth About Cell Phones and Cancer," *Scientific American*, October 4, 2010, http://www.scientificamerican.com/article.cfm?id=can-you-hear-me-now.

27. G. J. Rubin et al., "Idiopathic Environmental Intolerance Attributed to Electromagnetic Fields (Formerly 'Electromagnetic Hypersensitivity'): An Updated Systematic Review of Provocation Studies," *Bioelectromagnetics* 31, no. 1 (January 2010): 1–11, http://www.ncbi.nlm.nih.gov/pubmed/19681059.

28. Michael Grabell, "Europe Bans X-Ray Body Scanners Used at U.S. Airports," *ProPublica*, November 15, 2011, http://www.propublica.org/article/europe-bans-x-ray-body-scanners-used-at-u.s.-airports.

29. Richard Muller, *Physics for Future Presidents* (New York: Norton, 2008), 109.

30. Valerie Brown, "Is Radiation Actually Good for Some of Us?," Miller-McCune, February 17, 2012, http://www.miller-mccune.com/science/is-radiation-actually-good-for-some-of-us-39703/.

31. Nicola Nosengo, "Scientists Face Trial over Earthquake Deaths," *Nature News*, May 26, 2011, http://www.nature.com/news/2011/110526/full/news.2011.325.html.

32. Michael Day, "The Legal Aftershocks of the Earthquake in L'Aquila, Italy," *The Telegraph*, November 22, 2011, http://www.telegraph.co.uk/science/8905475/The-legal-aftershocks-of-the-earthquake-in-LAquila-Italy.html.

33. Charles Wheelan, *Naked Economics* (New York: Norton, 2010), 93.

34. Victoria Ward and Nick Collins, "EU Bans Claim That Water Can Prevent Dehydration," *The Telegraph*, November 18, 2011, http://www.telegraph.co.uk/news/worldnews/europe/eu/8897662/EU-bans-claim-that-water-can-prevent-dehydration.html.

35. "Where's Britain's Bill Gates?," *The Economist*, August 6, 2011, http://www.economist.com/node/21525406.

36. Martin Grueber and Tim Studt, "2011 Global R&D Funding Forecast," *R&D Magazine*, December 15, 2010, http://www.battelle.org/aboutus/rd/2011.pdf.

37. National Science Board, *Science and Engineering Indicators 2010* (Arlington, VA: National Science Foundation, 2010), http://www.nsf.gov/statistics/seind10/c/cs1.htm.

38. Ronald Reagan, "Radio Address to the Nation on the Federal Role in Scientific Research," American Presidency Project, April 2, 1988, http://www.presidency.ucsb.edu/ws/?pid=35637#axzz1hJWfwZBA.

39. See National Science Board, *Science and Engineering Indicators*.

40. Jocelyn Kaiser, "NIH Budget Reaches Doubling Point," *Science Now*, January 25, 2002, http://news.sciencemag.org/sciencenow/2002/01/25–04.html.

41. Peter Gosselin and Maura Reynolds, "After a Decade of Cuts, NASA to Get a Budget Boost," *Los Angeles Times*, February 4, 2003, http://articles.latimes.com/2003/feb/04/nation/na-bush4.

42. Martin Grueber and Tim Studt, "2012 Global R&D Funding Forecast," *R&D Magazine*, December 15, 2010, http://www.battelle.org/aboutus/rd/2012.pdf.

43. Ibid.

44. Ibid.

45. Ibid.

46. David Shukman, "China 'to Overtake US on Science' in Two Years," BBC, March 28, 2011, http://www.bbc.co.uk/news/science-environment-12885271.

47. David Cyranoski, "Strong Medicine for China's Journals," *Nature News*, September 15, 2010, http://www.nature.com/news/2010/100915/full/467261a.html.

48. Anna Greenspan, "China vs. America? Learning Strategies in the 21st Century," *The Globalist*, August 25, 2008, http://www.theglobalist.com/storyid.aspx?StoryId=5264.

49. "World's Best Universities: Top 400," *U.S. News & World Report*, 2010, http://www.usnews.com/education/worlds-best-universities-rankings/top-400-universities-in-the-world.

50. John Steele Gordon, "English: The Inescapable Language," *The American*, August 18, 2011, http://www.american.com/archive/2011/august/english-the-inescapable-language.

51. Ross Pomeroy, "China Is More Science Illiterate than the U.S., for Now," RealClearScience Newton Blog, October 11, 2011, http://www.realclearscience.com/blog/2011/10/if-you-think-the-us-has-science-problems-look-at-china.html.

Chapter 10

1. Lawrence Summers, "Remarks at NBER Conference on Diversifying the Science and Engineering Workforce" (transcript), Harvard University, January 14, 2005, http://president.harvard.edu/speeches/summers_2005/nber.php.

2. David Usborne, "Summers' 'Sexism' Costs Him Top Treasury Job," *The Independent*, November 24, 2008, http://www.independent.co.uk/news/world/americas /summers-sexism-costs-him-top-treasury-job-1033373.html.

3. William Saletan, "Sex on the Brain," *Slate*, November 17, 2011, http://www.slate.com/articles/health_and_science/human_nature/2011/11/boys_brains _girls_brains_how_to_think_about_sex_differences_in_psychology_.html.

4. Ibid.

5. See Summers, "Remarks at NBER Conference."

6. Mark J. Perry, "Shame on UC Davis and Reporter Sharon Stello," Carpe Diem, September 16, 2007, http://mjperry.blogspot.com/2007/09/shame-on-uc-davis-and -reporter-sharon.html.

7. Jeanna Bryner, "Men Smarter Than Women, Scientist Claims," *Live Science*, September 8, 2006, http://www.livescience.com/7154-men-smarter-women-scientist -claims.html.

8. "Intelligence in Men and Women Is a Gray and White Matter," *Science Daily*, January 22, 2005, http://www.sciencedaily.com/releases/2005/01/050121100142.htm.

9. Association for Psychological Science, "Women's Choices, Not Abilities, Keep Them Out of Math-Intensive Fields," press release, October 26, 2010, http://www .psychologicalscience.org/index.php/news/releases/womens-choices-not-abilities-keep -them-out-of-math-intensive-fields.html.

10. "Mattel Says It Erred; Teen Talk Barbie Turns Silent on Math," *New York Times*, October 21, 1992, http://www.nytimes.com/1992/10/21/business/company-news-mattel-says-it-erred-teen-talk-barbie-turns-silent-on-math.html.

11. "Balls and Brains," *The Economist*, December 4, 2008, http://www.economist .com/node/12719355.

12. Profile of Esteban González Burchard, MD, MPH, University of California–San Francisco, December 15, 2010, http://pulmonary.ucsf.edu/faculty/burchard.html.

13. Scott Freeman and Jon C. Herron, *Evolutionary Analysis*, 3rd ed. (Upper Saddle River, NJ: Pearson Prentice Hall, 2004), 321–325.

14. M. J. Dunn and R. Searle, "Effect of Manipulated Prestige-Car Ownership on Both Sex Attractiveness Ratings," *British Journal of Psychology* 101 (2010): 69–80, http://onlinelibrary.wiley.com/doi/10.1348/000712609X417319/abstract.

15. Jonathan Leake, "Wealthy Men Give Women More Orgasms," *Sunday Times*, January 18, 2009, http://www.timesonline.co.uk/tol/news/science/article5537017.ece.

16. Jennifer Abbasi, "Controversial Ideas: Does Semen Act as an Antidepressant to the Recipient?," *Popular Science*, May 4, 2011, http://www.popsci.com/science /article/2011–05/controversial-ideas-semen-natures-antidepressant.

17. Ibid.

18. Andy Bloxham, "Chivalry Is Actually 'Benevolent Sexism,' Feminists Conclude," *The Telegraph*, June 15, 2011, http://www.telegraph.co.uk/relationships /8575363/Chivalry-is-actually-benevolent-sexism-feminists-conclude.html.

19. J. C. Becker and J. K. Swim, "Seeing the Unseen: Attention to Daily Encounters with Sexism as Way to Reduce Sexist Beliefs," *Psychology of Women Quarterly* 35, no. 2 (2011): 227–242.

20. See http://www.sagepub.com/journals/Journal202010.

21. See Becker and Swim, "Seeing the Unseen."

22. Penn State University, "Sex Hormones Impact Career Choices," press release, September 1, 2011, http://live.psu.edu/story/54825.

23. Ibid.

24. Paola Sapienza et al., "Gender Differences in Financial Risk Aversion and Career Choices Are Affected by Testosterone," *Proceedings of the National Academy of Sciences* 106, no. 36 (2009): 15268–15273, http://www.ncbi.nlm.nih.gov/pmc/articles/PMC2741240/.

25. Tom Jacobs, "On 'Jeopardy!,' Women Take Fewer Risks vs. Men," Miller-McCune, June 2, 2011, http://www.miller-mccune.com/culture-society/on-jeopardy-women-take-fewer-risks-vs-men-31797/.

26. Coco Masters, "Study: Why Girls Like Pink," *Time*, August 20, 2007, http://www.time.com/time/health/article/0,8599,1654371,00.html.

27. Ian Sample, "Infant Chimps Play with 'Stick Dolls,'" *The Guardian*, December 20, 2010,
http://www.guardian.co.uk/science/2010/dec/20/chimps-play-male-female-genetic.

28. Steve Connor, "Genetic Tests Prove the 'Fairer Sex' Is Kinder Too," *The Independent*, February 9, 2011, http://www.independent.co.uk/news/science/genetic-tests-prove-the-fairer-sex-is-kinder-too-2208515.html.

29. Stephanie Pappas, "Stress Brings Out the Difference in Male, Female Brains," *Live Science*, September 29, 2010, http://www.livescience.com/10140-stress-brings-difference-male-female-brains.html.

30. University of California–San Francisco, "Men and Women's Immune Systems Respond Differently to PTSD," press release, April 26, 2011, http://www.ucsf.edu/news/2011/04/9790/mens-and-womens-immune-systems-respond-differently-ptsd.

31. Laura Sanders, "Cocaine Trumps Food for Female Rats," *Science News*, December 4, 2010, http://www.sciencenews.org/view/generic/id/65536/title/Cocaine_trumps_food_for_female_rats

32. "Feminist Perspectives on Sex and Gender," *Stanford Encyclopedia of Philosophy*, November 21, 2011, http://plato.stanford.edu/entries/feminism-gender/.

33. Ross Urken, "Let the Arguments Fly: Study Shows Women More Likely to Cause Traffic Accidents," *AOL Autos*, July 6, 2011, http://autos.aol.com/article/women-worse-drivers/.

34. Stuart Wolpert, "Psychologists Report That a Gender Gap in Spatial Skills Starts in Infancy," UCLA, December 7, 2008, http://newsroom.ucla.edu/portal/ucla/psychologists-report-a-gender-72612.aspx.

35. Seth Borenstein, "Women Drivers? They're Safer Than Men," MSNBC, January 20, 2007, http://www.msnbc.msn.com/id/16698153/ns/technology_and_science-science/t/women-drivers-theyre-safer-men/.

36. Rikki King, "Women's Short Shorts Make Multiple-Vehicle Crash on I-5 Worse," *The Herald*, August 18, 2010, http://www.heraldnet.com/article/20100818/NEWS01/708189659.

Chapter 11

1. U.S. Census, accessed March 13, 2012, http://quickfacts.census.gov/qfd/states/00000.html.

2. California Institute of Technology, Office of the Registrar, accessed January 12, 2012, http://www.registrar.caltech.edu/statistics.htm.

3. Jessica Pellien, "Is There A Bias Against College Applications from Asian Students?," Princeton University Press Blog, http://press.princeton.edu/blog/2011/12/05/is-there-a-bias-against-college-applications-from-asian-students/.

4. Jesse Washington, "Some Asians' College Strategy: Don't Check 'Asian,'" Associated Press, December 4, 2011.

5. Ralph K. M. Haurwitz, "Appeals Court Lets Race-Based Admissions at UT Stand," *American Statesman*, June 21, 2011.

6. National Science Foundation, "America's Pressing Challenge—Building a Stronger Foundation," February 2006, http://www.nsf.gov/statistics/nsb0602/.

7. Pascal D. Forgione, Jr., PhD, hearing on "Are Our Children Ready to Learn?," United States Senate Committee on Labor and Human Resources, December 4, 1998, http://nces.ed.gov/Pressrelease/senhrtest.asp.

8. DARPA-RA-10–03 Computer Science—Science, Technology, Engineering, and Mathematics (CS-STEM) Education Research Announcement (RA), https://www.fbo.gov/utils/view?id=69c81b4b7f892d4e0e0d8a7bec0eba29.

9. Christine M. Matthews, "Foreign Science and Engineering Presence in U.S. Institutions and the Labor Force" (Washington, DC: Congressional Research Service, October 28, 2010).

10. Thomson-Reuters ScienceWatch.com, November 2010, http://sciencewatch.com/dr/cou/2010/.

11. The White House, Office of the Press Secretary, Remarks by the President in the State of the Union Address, January 25, 2011, http://www.whitehouse.gov/the-press-office/2011/01/25/remarks-president-state-union-address.

12. Jon Bruner, "American Leadership in Science, Measured in Nobel Prizes," *Forbes*, October 5, 2011.

13. George Lakoff, *Don't Think of an Elephant!: The Essential Guide for Progressives.* (White River Junction, VT: Chelsea Green, 2004), 3–4.

14. George Lakoff, *Moral Politics: How Liberals and Conservatives Think* (Chicago: University of Chicago Press, 2002), 420.

15. Jerrold Meinwald and John G. Hildebrand, "Science and the Educated American: A Core Component of Liberal Education," American Academy of Arts and Sciences, Cambridge, MA, 2010, http://www.amacad.org/pdfs/SLACweb.pdf; International Center for the Advancement of Scientific Literacy, April 2011, http://icasl.org.

16. J. S. Hyde and J. Mertz, "Gender, Culture, and Math," *PNAS* 106 (2009): 8801–8807.

17. David Angus and Jeffrey Mirel, *The Failed Promise of the American High School, 1890–1995* (New York: Teachers College Press, 1999).

18. Aldous Huxley, *Brave New World* (New York: Doubleday), 19.

19. U.S. Department of Education, "No Child Left Behind Act Is Working," December 2006, http://www2.ed.gov/nclb/overview/importance/nclbworking.pdf.

20. Diane Ravitch, *The Death and Life of the Great American School System: How Testing and Choice Are Undermining Education* (New York: Basic Books, 2010), 21.

21. Tom Loveless, "The 2010 Brown Center Report on American Education," no. 13 (Washington, DC: Brookings Institution, February 7, 2011).

22. See http://stats.oecd.org/PISA2009Profiles/.

23. Sarah Meik, "The Chinese Teacher's Paycheck," AbroadChina.org, July 16, 2009, http://www.abroadchina.org/salary.asp.

24. Andrew J. Rotherham, "Shanghai Surprise: Don't Sweat Global Test Data," *Time*, January 20, 2011.

25. John Hechinger and Esmé E. Deprez, "Gates Urges Revamp of Teachers' $59 Billion Pay Plans," Bloomberg.com, November 19, 2010.

26. Sam Dillon, "Gates Urges School Budget Overhauls," *New York Times*, November 19, 2010.

27. Joy Resmovits, "Arne Duncan Boosts Merit Pay at Teaching Conference," Huffington Post, July 29, 2011.

28. Wynne Parry, "Educators Applaud Obama's Push for Math, Science Teaching," *Live Science*, January 26, 2011.

29. John S. Partington, "H. G. Wells: A Political Life," *Utopian Studies* 19 (2008): 517–576.

30. The Academic Competitiveness Council (ACC) report on STEM program's 2010 budget documents funding levels for each program specified by the ACC study.

31. The White House, Office of the Press Secretary, Transcript of the President's State of the Union Address, January 25, 2011, http://whitehouse.gov/the-press-office/2011/01/25/remarks-president-state-union-address.

32. Laura Meckler, "Education Push Includes Merit Pay," *Wall Street Journal*, March 11, 2009.

33. U.S. Department of Education biography, http://www2.ed.gov/news/staff/bios/duncan.html.

34. John Hechinger, "U.S. Teens Lag as China Soars on International Test," Bloomberg.com, December 7, 2010.

35. Richard M. Ingersoll and David Perda, "Is the Supply of Mathematics and Science Teachers Sufficient?," *American Educational Research Journal* 47, no. 3 (September 2010): 563–594.

36. Jeffrey Mervis, "Data Say Retention Is Better Answer to 'Shortage' Than Recruitment," *Science*, October 29, 2010, 580–581.

37. Ibid.

38. Ibid.

39. Ibid.

40. American Federation of Teachers, "Reversing Course: The Troubled State of Academic Staffing and a Path Forward," October 2008.

41. National Science Foundation, "Numbers of Doctorates Awarded Continue to Grow in 2009; Indicators of Employment Outcomes Mixed," 11–305, November 2010, http://www.nsf.gov/statistics/infbrief/nsf11305/nsf11305.pdf; "The Disposable Academic," *The Economist*, December 16, 2010, http://www.economist.com/node/17723223.

42. See http://colleges.usnews.rankingsandreviews.com/best-colleges/stony-brook-university-suny-196097/overall-rankings, April 2011.

43. See http://www.kiplinger.com/tools/colleges/, April 2011.

44. Naomi Schaefer Riley, "Why Unions Hurt Higher Education," *USA Today*, March 3, 2011.

45. Naomi Schaefer Riley, *The Faculty Lounges and Other Reasons Why You Won't Get the College Education You Paid For* (Lanham, MD: Ivan R. Dee, 2001), 113; Gene Russo, "Outreach: Meet the Press," *Nature* 468 (2010): 465–467.

Chapter 12

1. David Rowan, "How to Save Science Journalism," *Wired*, September 27, 2010.

2. Aisling Irwin," Science Journalism 'Flourishing' in Developing World," SciDev.net, February 18, 2009, http://www.scidev.net/en/news/science-journalism-flourishing-in-developing-world.html.

3. See http://www.livescience.com/7094-study-polar-bear-genitals-shrinking.html.

4. Ben Goldacre, "Trading Ideas in a Toilet," Bad Science, February 18, 2009.

5. David Rowan, "How to Save Science Journalism," *Wired*, September 27, 2010.

6. Goldacre, "Trading Ideas in a Toilet."

7. Hank Campbell, "Science Journalists Have Met the Enemy, and They Are Bloggers," Science 2.0, February 21, 2010.

8. Neil Collins, "Global Warming Generates Hot Air," *The Telegraph*, May 16, 2005.

9. Fred Pearce, "Flooded Out," *New Scientist*, June 5, 1999, http://www.new scientist.com/article/mg16221893.000-flooded-out.html.

10. Gwynne Dyer, "'Glaciergate' May Be Deal-Breaker," *Winnipeg Free Press*, January 26, 2010, http://www.winnipegfreepress.com/opinion/columnists/glacier gate-may-be-deal-breaker-82678787.html.

11. Nick Davies, "Churnalism Has Taken the Place of What We Should Be Doing: Telling the Truth," *Press Gazette*, February 4, 2008.

12. "Potato-Heads," *The Economist*, August 13, 1998, http://www.economist.com /node/171778.

13. Martin Enserink, "The Lancet Scolded over Pusztai Paper," *Science* 286, no. 5440 (October 1999): 656, doi:10.1126/science.286.5440.656a.

14. Interview with Gary Taubes, Science 2.0, January 3, 2008.

15. "Americans Are Losing the Victory in Europe," *Life*, January 7, 1946, 23.

16. Rebecca Leung, "For the Record: Bush Documents," CBS News, February 11, 2009.

17. World Federation of Science Journalists, "Science Journalism: Good and Bad News," March 23, 2009, http://www.wfsj.org/news/news.php?id=149.

18. "Science Journalism in Decline," *Sandwalk*, March 19, 2009.

19. Gene Russo, "Outreach: Meet the Press," *Nature* 468 (2010): 465–467.

20. J. A. Kleypas et al., "Geochemical Consequences of Increased Atmospheric Carbon Dioxide on Coral Reefs," *Science* 284, no. 5411 (1999): 118–120.

Chapter 13

1. "About Science Progress," http://scienceprogress.org/about/.

2. Jonathan D. Moreno and Rick Weiss, "Time for Science to Reclaim Its Progressive Roots," Science Progress, March 11, 2009, http://scienceprogress.org/2009/03 /science-next-excerpt/.

3. Pew Research Center, "Public Praises Science; Scientists Fault Public, Media," July 9, 2009,
http://www.people-press.org/2009/07/09/section-5-evolution-climate-change-and -other-issues/.

4. Ibid.

5. Ron Bailey, "Who's More Anti-Science: Republicans or Democrats?," Reason, December 27, 2011,
http://reason.com/archives/2011/12/27/whos-more-anti-science-republicans-or-de.

6. Frank Newport, "Republicans, Democrats Differ on Creationism," Gallup, July 20, 2008, http://www.gallup.com/poll/108226/republicans-democrats-differ-creation ism.aspx.

7. Pew Research Center, "Public Praises Science."

8. "Republicans Are More Scientifically Literate Than Democrats or Independents Are," Audacious Epigone, March 26, 2011,
http://anepigone.blogspot.com/2011/03/republicans-are-more-scientifically.html.

9. Razib Khan, "The Republican Fluency with Science," *Discover*, March 28, 2011, http://blogs.discovermagazine.com/gnxp/2011/03/the-republican-fluency-with -science/.

10. Alex Berezow, "GOP Might Be Anti-Science, but So Are Democrats," *USA Today*, September 20, 2011,
   http://www.usatoday.com/news/opinion/forum/story/2011–09–20/gop-democrats -science-evolution-vaccine/50482856/News+-+Opinion%29.

11. Paul Raeburn, "Chris Mooney: Classic False Equivalence of the Political Abuse of Science," Knight Science Journalism Tracker, September 22, 2011,
   http://ksjtracker.mit.edu/2011/09/22/chris-mooney-classic-false-equivalence-of -the-political-abuse-of-science/.

12. K. C. Johnson, "Proving the Critics' Case," *Inside Higher Ed*, August 26, 2005, http://www.insidehighered.com/views/2005/08/26/johnson.

13. Michael Tomasky, "Can You Play False Equivalency!?," *The Guardian*, January 14, 2010, http://www.guardian.co.uk/commentisfree/michaeltomasky/2010/jan/14 /conservative-liberal-haiti-hurricane-katrina.

14. Joe Romm, "Classic False Equivalence on Political Abuse of Science," ClimateProgress, September 22, 2011, http://thinkprogress.org/romm/2011/09/22 /326556/classic-false-equivalence-on-political-abuse-of-science/.

15. Tom Jacobs, "Wording Change Softens Global Warming Skeptics," Miller -McCune, March 2, 2011, http://www.miller-mccune.com/science-environment /global-warming-skeptics-believe-in-climate-change-28772/#.

16. Raeburn, "Chris Mooney."

17. Kevin Drum "Why We Deny," *Mother Jones*, April 18, 2011, http://mother jones.com/kevin-drum/2011/04/why-we-deny.

18. "Public Praises Science."

19. Chris Mooney, "Political Science Abuse: The Crucial Role of 'Elites,'" Science Progress Action, October 11, 2011, http://scienceprogressaction.org/intersection /2011/10/political-science-abuse-the-crucial-role-of-elites/.

20. Paul M. Sniderman et al., "The Fallacy of Democratic Elitism: Elite Competition and Commitment to Civil Liberties," *British Journal of Political Science* 21, no. 3 (July 1991): 349–370.

21. Nick Wing, "Creationist Theme Park Supported by Democratic Kentucky Governor," Huffington Post, May 25, 2011, http://www.huffingtonpost.com/2010/12/01 /kentucky-creationist-theme-park_n_790283.html.

22. Bruce Wilson, "Joining GOP's Bold March Backwards, Bobby Jindal and Louisiana Democrats Pass 'Stealth Creationism' Education Bill," Huffington Post, June 27, 2008,
   http://www.huffingtonpost.com/bruce-wilson/joining-gops-bold-march -b_b_109595.html.

23. Committee on Identifying and Assessing Unintended Effects of Genetically Engineered Foods on Human Health and National Research Council, *Safety of Genetically Engineered Foods: Approaches to Assessing Unintended Health Effects* (Washington, DC: National Academies Press, 2004), 8, http://www.nap.edu/ openbook.php?record_id=10977&page=8.

24. "Genetically Engineered Crops Benefit Many Farmers, but the Technology Needs Proper Management to Remain Effective," News from the National Academies, April 13, 2010, http://www8.nationalacademies.org/onpinews/newsitem.aspx?Record ID=12804

25. Alison L Van Eenennaam and William M Muir, "Transgenic Salmon: A Final Leap to the Grocery Shelf?," *Nature Biotechnology* 29 (2011): 706–710, doi:10.1038/nbt.1938.

26. Center for Veterinary Medicine, "Environmental Assessment for AquAdvantage® Salmon," Food and Drug Administration August 25, 2010, http://www.fda.gov/downloads/AdvisoryCommittees/CommitteesMeetingMaterials/VeterinaryMedicineAdvisoryCommittee/UCM224760.pdf.

27. Michelle Theriault Boots, "Alaska Lawmakers Ask FDA to Quash Company's Bid to Grow Genetically Modified Salmon," KTUU-TV, July 15, 2011, http://articles.ktuu.com/2011–07–15/aquabounty-technologies_29779898.

28. Romm, "Classic False Equivalence."

29. Frangellica Angel, "Murkowski and Begich to FDA: Do Not Approve Frankenfish," *Alaska News*, July 27, 2011, http://thealaskanews.com/murkowski-begich-fda-approve-frankenfish/9949.

30. Rod Dreher, "Science vs. Religion: What Do Scientists Say?," Beliefnet, April 30, 2010, http://blog.beliefnet.com/roddreher/2010/04/science-vs-religion-what-do-scientists-say.html.

31. Neil deGrasse Tyson, "Neil deGrasse Tyson on Science and Faith," BigThink, February 19, 2009, http://bigthink.com/ideas/13148.

32. Roger Pielke Jr., "The New Eugenics from the Looney Left," Roger Pielke Jr.'s Blog, November 8, 2011, http://rogerpielkejr.blogspot.com/2011/11/new-eugenics-from-looney-left.html.

Chapter 14

1. Lord Rothschild HL Deb 10 April 1946 vol. 140 cc643–75.

2. Kate Lowenstein, "Not Safe to Eat: Three Foods to Avoid," CNN, October 12, 2011, http://www.cnn.com/2011/10/12/health/food-poisoning-protection-guide/index.html.

3. Quoted by Thomas C. Leonard, "Retrospectives: Eugenics and Economics in the Progressive Era," *Journal of Economic Perspectives* 19, no. 4 (Fall 2005): 207–224, http://www.princeton.edu/~tleonard/papers/retrospectives.pdf.

4. Jean H. Baker, *Margaret Sanger: A Life of Passion* (New York: Hill and Wang, 2011), 145.

5. Victoria Brignell, "The Eugenics Movement Britain Wants to Forget," *New Statesman*, December 9, 2010, http://www.newstatesman.com/society/2010/12/british-eugenics-disabled.

6. "Charlie Sheen—Winning," YouTube, March 1, 2011, http://www.youtube.com/watch?v=pipTwjwrQYQ.

7. J. C. Becker and J. K. Swim, "Seeing the Unseen: Attention to Daily Encounters with Sexism as a Way to Reduce Sexist Beliefs," *Psychology of Women Quarterly* 35 (2011): 227–242.

8. Jane Merriman, "After Gender Bias, Women Face Gender Fatigue," Reuters, November 2, 2009, http://www.reuters.com/article/2009/11/02/us-gender-fatigue-idUSTRE5A13HE20091102.

9. Gregory M. Walton and Steven J. Spencer, "Latent Ability: Grades and Test Scores Systematically Underestimate the Intellectual Ability of Negatively Stereotyped Students," *Psychological Science* 20, no. 9 (September 2009): 1132–1139.

10. See http://www.census.gov/hhes/www/eeoindex/page_c.html?.

11. Christopher F. Cardiff and Daniel B. Klein, "Faculty Partisan Affiliations in All Disciplines: A Voter Registration Study," *Critical Review* 17, nos. 3–4 (2005), http://www.criticalreview.com/2004/pdfs/cardiff_klein.pdf.

12. John Tierney, "Social Scientist Sees Bias Within," *New York Times*, February 7, 2011.

13. Stanley Rothman, S. Robert Lichter, and Neil Nevitte, "Politics and Professional Advancement Among College Faculty," *The Forum* 3, no. 1 (2005), http://www.cwu.edu/~manwellerm/academic%20bias.pdf.

14. Jerry Coyne, "Chris Mooney, Evolution, and Politics," February 9, 2012, http://whyevolutionistrue.wordpress.com/2012/02/09/chris-mooney-evolution-and-politics/.

Chapter 15

1. Nell Greenfieldboyce, "'Shrimp on a Treadmill': The Politics of 'Silly' Studies," NPR, August 23, 2011, http://www.npr.org/2011/08/23/139852035/shrimp-on-a-treadmill-the-politics-of-silly-studies.

2. "Thou Orb Aloft Full-Dazzling," *The Economist*, October 15, 2011, http://www.economist.com/node/21532285.

3. Michael Webber, "We Have a New Opportunity for Sensible Energy Policy," *Dallas Morning News*, November 27, 2006.

4. "Frack On," *The Economist*, November 26, 2011, http://www.economist.com/node/21540275.

5. Jon Entine, "Future Energy: Natural Gas Fracking—Who Blew Up the 'Bridge to the Future'?," American Enterprise Institute, December 13, 2011, http://www.aei.org/article/energy-and-the-environment/conventional-energy/natural-gas/who-blew-up-the-bridge-to-the-future/.

6. World Nuclear Association, "Nuclear Power in France," accessed February 3, 2012, http://www.world-nuclear.org/info/inf40.html.

7. Jason Mick, "New Solar Cell Gives Its '110 Percent' in Efficiency," *Daily Tech*, December 20, 2011, http://www.dailytech.com/New+Solar+Cell+Gives+Its+110+Percent+in+Efficiency/article23548.htm.

8. Rebecca Boyle, "The Moon Should Be the 51st State, and Other Space Dreams from Newt Gingrich," *Popular Science*, January 26, 2012, http://www.popsci.com/science/article/2012–01/gingrich-moon-should-be-51st-state.

9. Clara Moskowitz, "NASA in Transition as Congress OKs New Direction," Space.com, September 30, 2010, http://www.space.com/9233-nasa-transition-congress-oks-direction.html.

10. Laura Beil, "What Happens to Extra Embryos After IVF?," CNN, September 1, 2009, http://www-cgi.cnn.com/2009/HEALTH/09/01/extra.ivf.embryos/index.html.

11. Sumit Paul-Choudhury, "Digital Legacy: The Fate of Your Online Soul," *New-Scientist*, May 2, 2011, http://www.newscientist.com/article/mg21028091.400-digital-legacy-the-fate-of-your-online-soul.html?full=true.

12. "Mind-Goggling," *The Economist*, October 29, 2011, http://www.economist.com/node/21534748.

13. F. C. Li et al., "Finding the Real Case-Fatality Rate of H5N1 Avian Influenza," *Journal of Epidemiology and Community Health* 62, no. 6 (2008): 555–559, http://www.ncbi.nlm.nih.gov/pubmed/18477756.

14. "Newly Engineered Highly Transmissible H5N1 Strain Ignites Controversy About Balancing Scientific Discovery and Public Safety," *Science Codex*, January 26,

2012, http://www.sciencecodex.com/read/newly_engineered_highly_transmissible
_h5n1_strain_ignites_controversy_about_balancing_scientific_discovery_and_public
_safet.

15. Tia Ghose, "The Risks of Dangerous Research," *The Scientist*, January 13, 2012, http://the-scientist.com/2012/01/13/the-risks-of-dangerous-research/.

16. Phillip Matier and Andrew Ross, "SF Animal Control Commission Seeks Ban on Goldfish," *San Francisco Chronicle*, June 15, 2011, http://www.sfgate.com/cgi-bin/article.cgi?f=/c/a/2011/06/14/BA661JTO52.DTL#ixzz1PMXsmDnR.

17. Frances Martel, "John Stossel's 'Wackiest Warning Labels' Are All Very Dumb," Mediaite, June 18, 2011, http://www.mediaite.com/tv/john-stossels-wackiest-warning-labels-are-all-very-dumb/.

18. "Over-Regulated America," *The Economist*, February 18, 2012, http://www.economist.com/node/21547789.

19. Henry Miller, "How Is the FDA Really Doing?," *Genetic Engineering and Biotechnology News*, February 15, 2012, http://www.genengnews.com/gen-articles/how-is-the-fda-really-doing/3999/.

20. Ryan Flinn, "Venture Firms Reduce Biotechnology Investment on FDA Risk," Bloomberg, October 6, 2011, http://www.bloomberg.com/news/2011–10–06/venture-firms-pull-back-biotechnology-investment-on-fda-hurdles.html.

21. Sydney Spiesel, "Sick City," *Slate*, September 6, 2005, http://www.slate.com/articles/health_and_science/medical_examiner/2005/09/sick_city.html.

22. Alex Berezow, "We Must Irradiate Food Supply," RealClearScience Newton Blog, August 17, 2011, http://www.realclearscience.com/blog/2011/08/we-must-irradiate-food-supply.html.

23. Brian Wallheimer, "*E. coli, Salmonella* May Lurk in Unwashable Places in Produce," Purdue University, August 15, 2011, http://www.purdue.edu/newsroom/research/2011/110815DeeringPathogens.html.

24. Miriam Falco, "CDC: 1 in 6 Americans Get Food Poisoning Annually," CNN, December 15, 2010, http://thechart.blogs.cnn.com/2010/12/15/cdc-1-in-6-americans-get-food-poisoning-annually/.

25. Alison Flood, "Scientists Sign Petition to Boycott Academic Publisher Elsevier," *The Guardian*, February 2, 2012, http://www.guardian.co.uk/science/2012/feb/02/academics-boycott-publisher-elsevier.

26. "The Price of Information," *The Economist*, February 4, 2012, http://www.economist.com/node/21545974.

27. David Dobbs, "Congress Considers Paywalling Science You Already Paid For," *Wired*, January 6, 2012, http://www.wired.com/wiredscience/2012/01/congress-considers-paywalling-science-you-already-paid-for/.

28. John Timmer, "Appeals Court Overrules Lower Court, Upholds Breast Cancer Gene Test," *Ars Technica*, July 2011, http://arstechnica.com/science/news/2011/07/appeals-court-overrules-lower-court-upholds-breast-cancer-gene-test.ars.

29. Associated Press Business Staff, "Supreme Court Throws Out Human Gene Patents," Cleveland.com, March 26, 2012, http://www.cleveland.com/business/index.ssf/2012/03/supreme_court_throws_out_human.html.

30. Alex Berezow, "The World Is Not Overpopulated," RealClearScience, July 20, 2011, http://www.realclearscience.com/articles/2011/07/20/the_world_is_not_overpopulated_106247.html.

31. Fred Pearce, "Consumption Dwarfs Population as Main Environmental Threat," *The Guardian*, April 15, 2009, http://www.guardian.co.uk/environment/2009/apr/15/consumption-versus-population-environmental-impact.

32. Ronald Bailey, "Deconsumption Versus Dematerialization," Reason, February 15, 2011, http://reason.com/archives/2011/02/15/deconsumption-versus-demateria.

33. Ibid.

34. Bryan Walsh, "The Worst Kind of Poverty: Energy Poverty," *Time*, October 11, 2011, http://www.time.com/time/health/article/0,8599,2096602,00.html.

35. Ibid.

36. World Health Organization, *The Global Burden of Disease: 2004 Update* (Geneva: World Health Organization, 2008), http://www.who.int/healthinfo/global_burden_disease/2004_report_update/en/index.html.

37. Brandon Keim, "Bush Triples Funding for AIDS, Malaria, and Tuberculosis," *Wired*, July 31, 2008, http://www.wired.com/wiredscience/2008/07/bush-triples-fu/.

38. "Hot Tropic," *The Economist*, February 4, 2012, http://www.economist.com/node/21546005.

39. "Defending Science," *The Economist*, January 31, 2002, http://www.economist.com/node/965718.

40. Ross Pomeroy, "BEST May Be Best Study Yet on Climate Change," RealClear-Science Newton Blog, October 23, 2011, http://www.realclearscience.com/blog/2011/10/new-climate-study.html.

Conclusion

1. Christina Wilkie, "Rep. Hank Johnson: Guam Could 'Tip over and Capsize,'" *The Hill*, March 31, 2010, http://washingtonscene.thehill.com/in-the-know/36-news/3169-rep-hank-johnson-guam-could-tip-over-and-capsize.

INDEX

Dr. Alex B. Berezow is the editor of RealClearScience. His work has appeared on CNN and in *USA Today* and *Forbes*, among other publications. In 2010, he earned a PhD in microbiology from the University of Washington. He lives with his wife in Seattle.

Hank Campbell is the founder of Science 2.0, the world's largest independent science communication community. Prior to founding Science 2.0 in 2006, he had a fifteen-year career as a senior executive in physics software, including one company that resulted in an IPO. He graduated from Duquesne University in 1987 with a degree in psychology and journalism and was formerly a U.S. Army officer. He lives in Sacramento.

PublicAffairs is a publishing house founded in 1997. It is a tribute to the standards, values, and flair of three persons who have served as mentors to countless reporters, writers, editors, and book people of all kinds, including me.

I. F. STONE, proprietor of *I. F. Stone's Weekly*, combined a commitment to the First Amendment with entrepreneurial zeal and reporting skill and became one of the great independent journalists in American history. At the age of eighty, Izzy published *The Trial of Socrates*, which was a national bestseller. He wrote the book after he taught himself ancient Greek.

BENJAMIN C. BRADLEE was for nearly thirty years the charismatic editorial leader of *The Washington Post*. It was Ben who gave the *Post* the range and courage to pursue such historic issues as Watergate. He supported his reporters with a tenacity that made them fearless and it is no accident that so many became authors of influential, best-selling books.

ROBERT L. BERNSTEIN, the chief executive of Random House for more than a quarter century, guided one of the nation's premier publishing houses. Bob was personally responsible for many books of political dissent and argument that challenged tyranny around the globe. He is also the founder and longtime chair of Human Rights Watch, one of the most respected human rights organizations in the world.

·   ·   ·

For fifty years, the banner of Public Affairs Press was carried by its owner Morris B. Schnapper, who published Gandhi, Nasser, Toynbee, Truman, and about 1,500 other authors. In 1983, Schnapper was described by *The Washington Post* as "a redoubtable gadfly." His legacy will endure in the books to come.

Peter Osnos, *Founder and Editor-at-Large*